U0224883

居住区景观设计

Community Landscape Design

（印）程奕智 / 编　常文心　杨莉 / 译

辽宁科学技术出版社

Contents 目录

Preface 前言

As the world is undergoing rapid urbanisation and demands for good housing in cities soar, we are looking for residential environments where we can take refuge from our intense and hectic lifestyle. This enticing book is a window to diverse range of community landscape design creating exceptional living environments across the globe addressing this need of high-quality housing. Filled with exciting images from projects designed by renowned international firms this book is a visual treat for all professionals, students and design lovers. Showcasing ways to choreograph unique experiences in residential developments it gives us a comprehensive portrait of inspiring landscape projects worldwide. This stimulating book is divided broadly into three sections – high-rise residential tower landscape, mid-rise residential block landscape and the low-rise villa landscape as each category has its unique landscape character. The high-rise residential blocks are increasingly integrating vertical green features as part of façade with sky decks and terrace gardens. Open spaces incorporating all the community facilities are generally on decks with parking below. In contrast, the landscape feel in mid-rise residential block developments is more of being in a courtyard environment. Very often roof gardens become another space for community interaction in this form of development as allotment gardens and urban farms. The landscape design for villa development with its low-rise architecture has its own intimate feel and character. With private gardens, leafy streets and community greens it is a delicate interplay of finer texture and scale evoking its unique charm and sense of place. Bringing all these disparate design approaches together under one comprehensive volume makes this book a stimulating and inspiring journey through the world of contemporary residential landscape.

Landscape design for residential developments is becoming increasingly diverse in its scale and ambition. Due to advancement of technology and blurring of work-life boundary, our home as a refuge is increasingly becoming crucial in fostering our well being. Hence designers are moving towards creating oasis like spaces where residential developments are conceived with lavish pools and extensive gardens. As a result, new outdoor experiences integrating art and culture are staring to evolve as the boundary between indoor and outdoor spaces blur. Landscape design is unique in many senses. The key tool for creating spaces remains the use of plants that grows and changes with time. The dynamic plant cycle with its colour and texture mixed with hardscape elements give landscape architects a vibrant set of materials to play with. Water with its various forms adds fluidity and reflection as well as calmness and movement. Lighting at night adds its own drama. Equipped with these tools landscape architects can create a unique and dramatic experiences in projects where culture, context, sustainability and budget form some of the key parameters.

The beautifully articulated projects in the book demonstrate how landscape architects integrate these parameters in their projects. For example, Shma's SUMMER project near Bangkok is a stunning choreography of art, architecture and nature. It is a landscape where context of sea is abstracted in a series of unique geometrical forms to create a dramatic experience. Linking culture and context in abstract form is demonstrated in the sublime design of Trop's Baan Sansuk development. Here the interplay between rocks and boulders with water create a grand setting for the residential courtyard. Another project that beautifully integrates the context of water in its physicality is Rolfsbukta in Oslo, Norway by Bjorbekk and Lindhelm where water is brought into the site through a canal connected to the bay. Sustainability plays a dominant role and becomes a big driver for projects like The Neo Bankside in London by Gillespies and Block 32 at Rino, Denver by studioINSITE. Both the projects generate unique character by using native planting and fostering local biodiversity. In Block 32 the installations integrated with solar panels as sculptural elements no doubt form a playful response to the project's location within the art district.

Creating compelling public spaces with limited budget for social housing projects is always a challenge. However projects like Exemplary Social Housing Project in Munich by Irene Burkhart Landscape Architects and 209 Guadabajaque Council Housing in Jerez de la Frontera, Spain by Acta Sim, Manuel Narvaez Perez and Fernando Visedo Manzanares show how simple and sensitive approach to design can create great community landscape for residents to enjoy. Inclusion of exemplary social housing landscape projects in this book is no coincidence as large population migrates to cities and look for affordable housing. Undoubtedly social housing landscape will be one of the key areas for future landscape design. Another new avenue of residential design will stem from the aging population. A design that is sensitive, imaginative and community focused will be the key in fulfilling user aspiration. Designers need to consider the aspects of safety, materiality and accessibility to generate a holistic living environment for the elderly.

In order to address global environmental issues as more regulation and certification will be required, landscape design as a whole will also evolve accordingly. One example will be the extensive use of roof gardens – spaces for both gardening and farming. We will increasingly see building façades, especially in high-rise buildings with more green walls in form of vertical greening. In a dense urban condition the roof gardens along with the green façades will evolve as elevated ecological network integrating landscape, architecture, art and engineering. As new ideas emerge and how we design community spaces evolve, our role as landscape architects will become more important than ever.

So watch the space!

Viraj Chatterjee

B Arch (Hons), MLA (Distinction), CMLI, MCA, AIIA
Founder and Design Principal, ONE landscape
Assistant Professor of Landscape Architecture, Hong Kong University

当今世界正经历着快速的城市化进程，而人们对优秀住宅项目的需求也节节攀升。我们都在寻找一种能够远离喧嚣、缓解压力的居住环境。本书作为一个窗口，向读者展示了来自全球各地的优秀居住区景观设计，这些景观设计对塑造高品质居住环境至关重要。本书收录了由国际知名景观事务所设计的居住区景观项目，精美的项目图片将为所有景观专业设计人员、学生和设计爱好者提供一场视觉盛宴。通过展示居住开发项目的独特设计流程，本书让读者对全球的优秀景观设计有一个综合的了解。这本极富启发性的图书分为三个部分：高层住宅景观、多层住宅景观以及别墅景观，每个类别都有独特的景观特色。高层住宅楼越来越多地将垂直绿化设计融入其中，在部分建筑外立面上设置了空中平台或露台花园。融入所有社区设施的开放空间通常都设在平台上，而下方是停车场。相反，多层住宅开发则更趋向于打造一种庭院环境。屋顶花园是另一种社区互动空间，经常以分户花园和城市农场的形式出现。由于别墅开发的建筑结构较低，它的景观设计有种独特的亲密感。私人花园、林荫街道和社区绿地穿插形成精致的纹理和比例，激发了别墅区的独特魅力和空间感。本书将这些风格各异的设计方案汇集起来，引领读者体验让人迸发灵感的全球现代居住区景观之旅。

居住区景观设计在尺度和形式上越来越多样化。随着科技的进步和工作与生活之间界限的模糊，家对保持我们的身体健康来说变得更加重要。因此，设计师开始在居住区中打造绿洲空间，添加丰富的水池和花园设施。文化和艺术相融合的新型户外体验开始演变，而室内外空间的边界也变得更加模糊。从多种意义来讲，景观设计都十分特殊。景观设计的主要工具是植物，它们会随着时间的流逝而成长变化。植物色彩和纹理的动态循环与硬景观元素相互融合，为景观设计师提供了丰富多变的设计材料。水以其千变万化的形式为景观添加了流畅感、倒影感、宁静感和运动感。夜晚的灯光富有戏剧效果。在文化、环境、可持续因素和项目预算等重要参数的引导下，这些工具让景观设计师得以创造出更独特的丰富体验。

本书所呈现的精美项目展示了景观设计师是如何将这些参数融入他们的项目之中的。例如，Shma 景观设计公司的夏日公寓项目（泰国，曼谷）出色地融合了艺术、建筑和自然。景观设计将海洋环境抽象成若干个独立几何造型，形成了生动的体验。TROP 公司的幸福之家项目（泰国，华欣）实现了文化与环境的抽象化连接。岩石与水的互动为住宅庭院营造出宏大的背景。由贝吉比克 & 林登景观事务所设计的罗尔夫斯湾项目（挪威，奥斯陆）同样巧妙地融入了水景元素，以景观运河的形式将居住区与海湾连接起来。可持续因素在景观设计中扮演了重要的角色。由吉尔斯派斯设计的尼奥河畔项目（英国，伦敦）和由 INSITE 工作室设计的北河 32 号街区项目（美国，丹佛）都在设计中采用了本土植物并促进了本地生物多样化，形成了独特的风格。北河 32 号街区项目将太阳能电池板装置设计成雕塑型元素，充分反映了项目所在的文化区特征。

如何用有限的预算打造出色的社会福利住房项目一直是设计师面临的重大挑战。由艾琳·布克哈特景观建筑与城市规划公司设计的示范性社会福利住房项目（德国，慕尼黑）和由阿克塔·斯利普、曼纽尔·纳尔瓦埃斯·佩雷斯、费尔南多·维斯多·曼萨纳雷斯三人合作设计的 209 古达巴佳克公寓（西班牙，赫雷斯）展示了如何利用简单有效的设计来塑造宜人的居住区景观。另一个居住区景观设计的新形式来自于人口的老龄化。只有富有针对性、创造力并且能聚焦社区的设计才能满足用户的渴望。设计师需要全面考虑安全性、物质性以及可达性，为老年人设计一个完整的居住环境。

为了应对全球环境问题，各国出台了更多与建筑相关的法律法规和资格认证，景观设计也必须随之演变。屋顶花园就是一个典型的例子，它能提供更多的花园和农作空间。我们将会发现越来越多的高层建筑在外墙上运用垂直绿化元素。在密集的城市环境中，屋顶花园和绿色墙壁将演化为空中生态网络，将景观、建筑、艺术和工程整合起来。随着新观念的出现和居住空间设计方式的演变，景观设计师的角色将会变得空前重要。

让我们关注空间，拭目以待。

程奕智

（建筑学学士、景观设计学硕士、注册景观设计师、注册建筑师、印度建筑师协会会员）
ONE 景观设计公司创始人兼设计总监
香港大学景观设计学助理教授

• *Mid-rise residential block landscape*

多层居住区景观

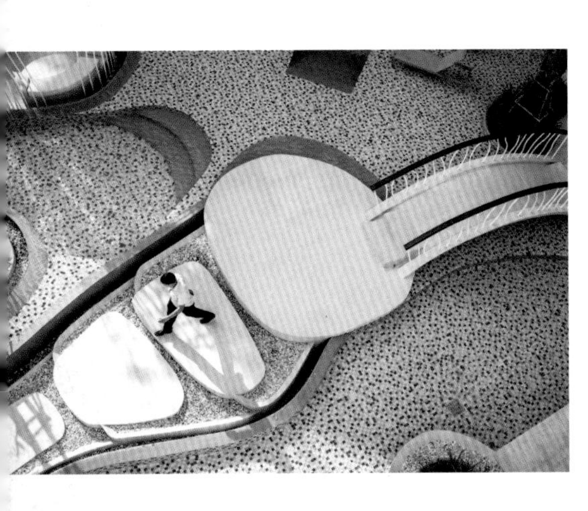

SUMMER
夏日公寓

Location: Bangkok, Thailand
Completion: 2013
Design: Shma Company Limited
Photography: Mr. Pirak Anurakyawachon
Area: 9,982 sqm
项目地点：泰国，曼谷
完成时间：2013年
设计师：Shma景观设计公司
摄影：皮拉克·阿努拉克亚瓦秋
面积：9,982平方米

The project is located at famous seaside town called Hua Hin which is situated about 300 km south of Bangkok. The site is in irregular shape of approximately 9,982 square metres. The main landscape area is surrounded by an L shape building where it leaves the other 2 sides visually connected with the nearby private garden. Since the site is situated far away from the beach and half of the residential units cannot receive sea view, the landscape design strategy is to create landscape that not only functions as an outdoor hideaway space but also reflects "Seascape" atmosphere as the residents would feel closer to the sea.

In other words, the design concept derives from the "sea bubble" which occurs once the wave roll over the shores. The bubbles come in various size and none of them are looking the same. The designers conceptually use the bubble form to generate the overall design. The landscape space is divided into two: swimming pool and garden. The pool is located at the higher level next to the building while garden is next to the wall next to private villas. The large water body of the pool covers 50% of the landscape area and spans across the whole length of building edge, providing a direct pool access of all ground floor units. It also contains various water related activities within this surface from 25 m lap pool, children pool, shallow pool, submerge pool deck with bubble water jet and Jacuzzi, to sunken cabana. With this unique curved shape of pool edge, it provides not only the ultimate setting for each activities but also a better functioning space for garden level which is lowered by 450 mm. The overflowing water at the pool edge creates a tranquil fluid sound ambience for the whole landscape area. At the garden, bubble shape becomes a primary form in generating spaces which composed of stepping stone, planters, lawn, gravel beds and cabanas.

There are 3 cabanas in cylindrical shape range in various heights from 4 to 5 metres which are nicely placed between the pool curvatures and become main focal points from all directions. One of the cabana is the pavilion to provide certain privacy for Jacuzzi zone while the other 2 are daybed and seating area which are only accessible from the garden level. The shape of the cabana and the pattern of the trellis are also inspired from "Sea Wave". A series of 30 mm diameter of linear tubes

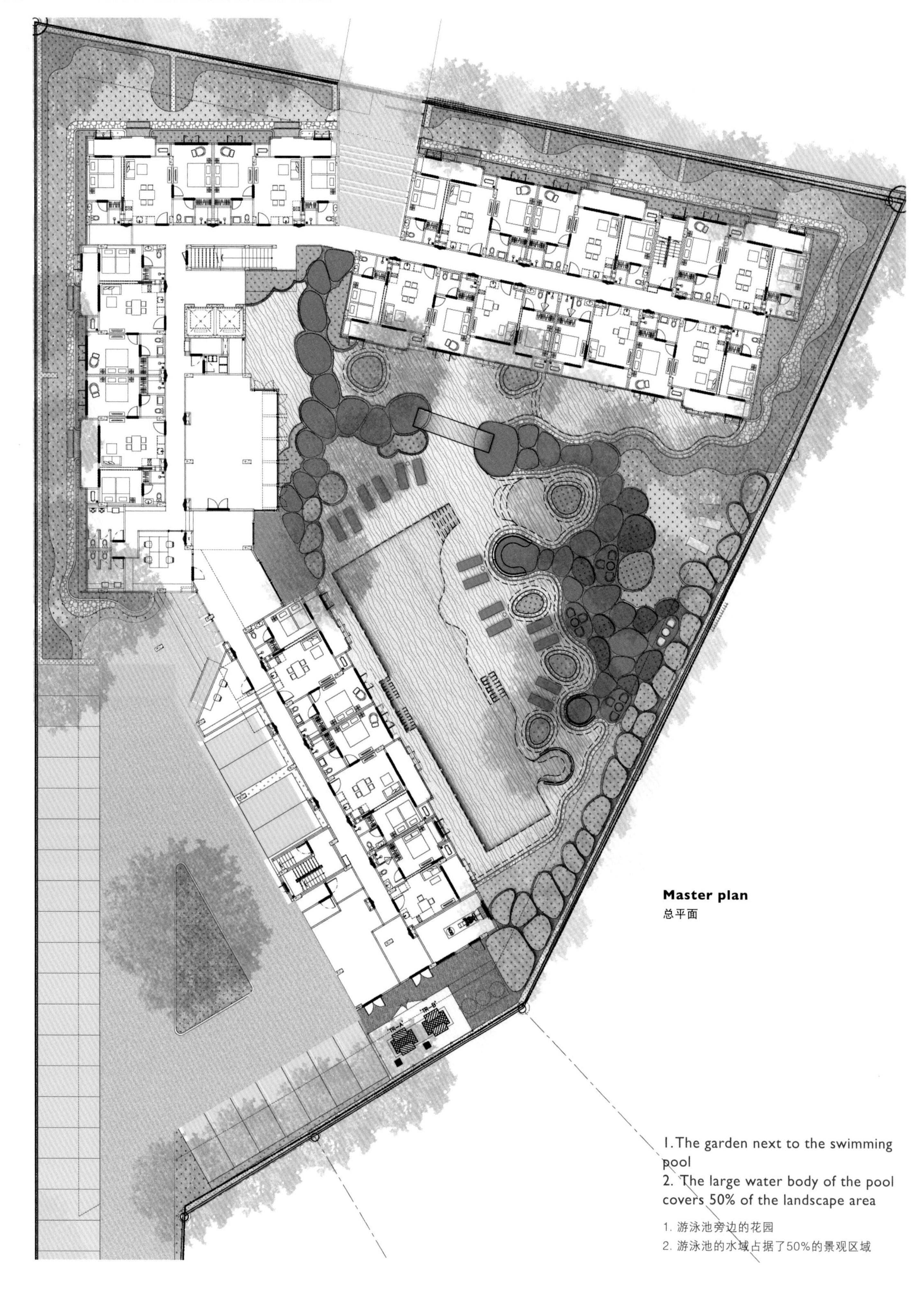

Master plan
总平面

1. The garden next to the swimming pool
2. The large water body of the pool covers 50% of the landscape area

1. 游泳池旁边的花园
2. 游泳池的水域占据了50%的景观区域

were bended to create curve lines at difference heights. As a result, the subtle rhythm of cabana shells exposes a unique light and shadow play. With its forms and white coloured structure of Trellis, it contrasts with dark green backdrop of the garden behind so it emphasises itself as a sculpture during the day and lanterns at night, with series of up lights that were integrated at each steel tubes.

The colour palette of hard-scape and soft-cape elements here is in neutral and light pastel colour composition. It is to create a calm and serene atmosphere for this hideaway place while giving a wide open feeling to the overall space.

This landscape space is the designer's interpretation of the "Seascape" for the hideaway of modern living.

项目位于泰国曼谷以南 300 公里的著名海滨小镇华欣。项目场地为 9,982 平方米的不规则形状，主要景观区两面由一座 L 形建筑环绕，另两面则与附近的私家花园连接起来。由于项目场地远离海滩并且有一半的住户无法享受海景，设计策略试图让景观不仅起到户外休闲空间的作用，同时还要营造出一种海景氛围，让住户感到更贴近大海。

换言之，设计的灵感来自于海浪打在沙滩上所形成的“海洋泡沫”。它们造型各异，没有任何两个泡沫是完全相同的。设计师运用泡沫概念完成了整个设计。景观空间一分为二：游泳池和花园。游泳池位于靠近建筑的高处，而花园则紧邻花园别墅的围墙。游泳池的水域占据了 50% 的景观区域，横跨真个建筑边缘的长度，让一楼住户享有直接进入泳池的便利。游泳池还包含各种各样与水相关的活动：25 米标准泳池、儿童泳池、浅水池、带有泡泡喷水口的浸入式泳池平台、极可意按摩浴缸以及下沉小屋。独特的弧形泳池边界让游泳池不仅为每项互动都提供了极致的背景，还为低处的花园区（地势低 450 毫米）营造了更好的功能空间。泳池边缘溢出的水发出涓涓水声，为整个景观区域营造出宁静的氛围。花园主要以泡沫形状进行空间划分，包含垫脚石、花坛、草坪、碎石带、凉亭小屋等空间类型。

景观区有三个圆柱形的凉亭小屋，高度分别在 4~5 米之间，它们巧妙地契合了游泳池的弧度，从各个方向看都是焦点。其中一个凉亭设在按摩浴缸旁，一定程度上保护了隐私，另两个则提供了坐卧两用座椅，仅可从花园进入。小屋的造型和棚架样式同样从“海浪”中获得了启发。直径为 30 毫米的长管被弯成不同高度的曲线造型。小屋外壳微妙的韵律感为内外空间带来了独特的光影效果。独特的造型和白色结构让棚架与深绿色的花园背景形成了对比。白天，它像一座雕塑；夜晚，钢管中的上照灯让它变成了一盏灯笼。

软硬景观元素的色调以黑白和清新的浅色为主，营造出恬静平和的隐居氛围，同时又让整个空间看起来更加开敞。

整个景观空间的设计就是设计师对现代生活隐居空间“海景”的诠释。

2

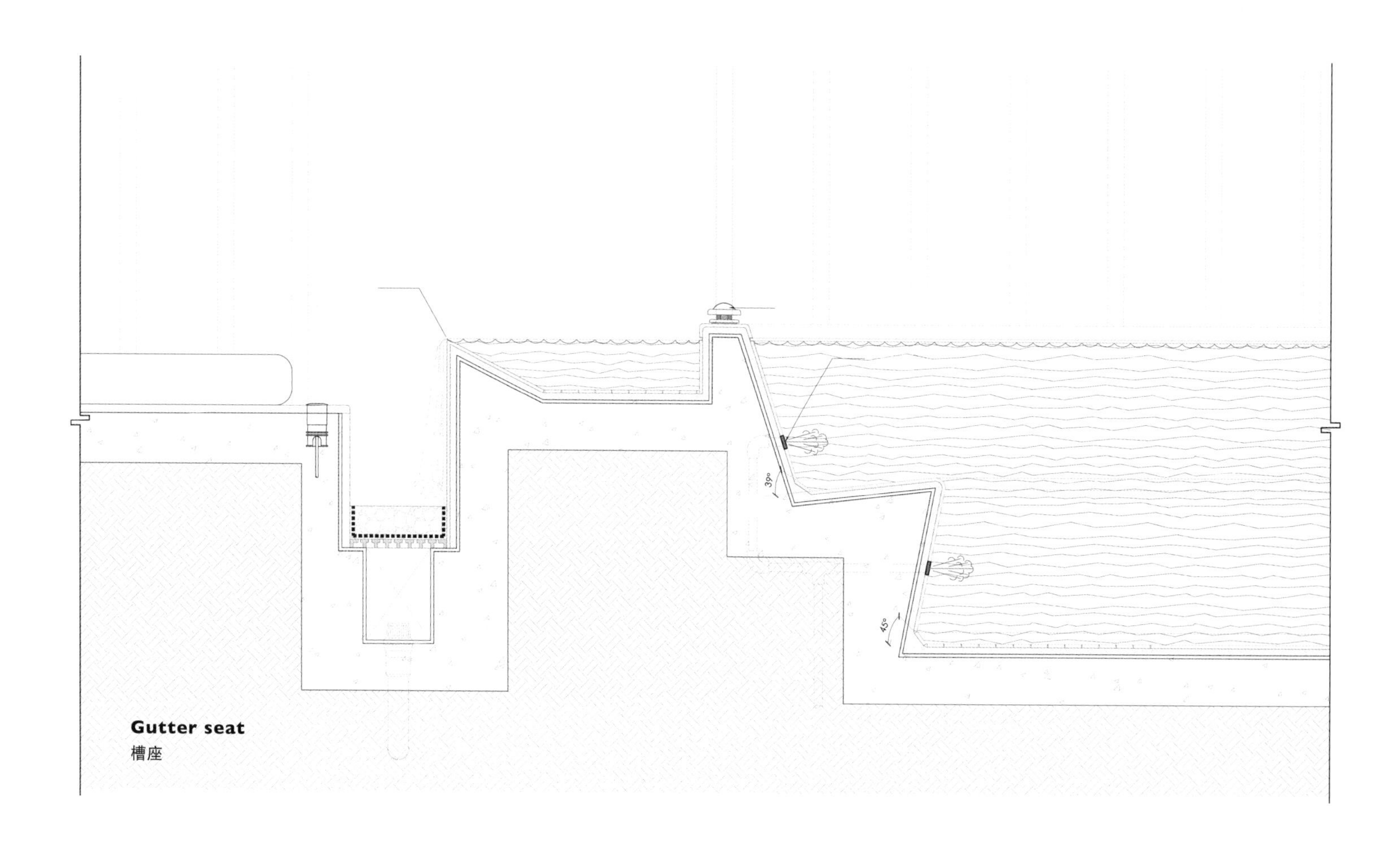

Gutter seat

槽座

3. A small bridge across the pool
4. Cabanas in cylindrical shape
5. The seating

1. 一座小桥跨于泳池之上
2. 圆柱形的凉亭小屋
3. 座椅

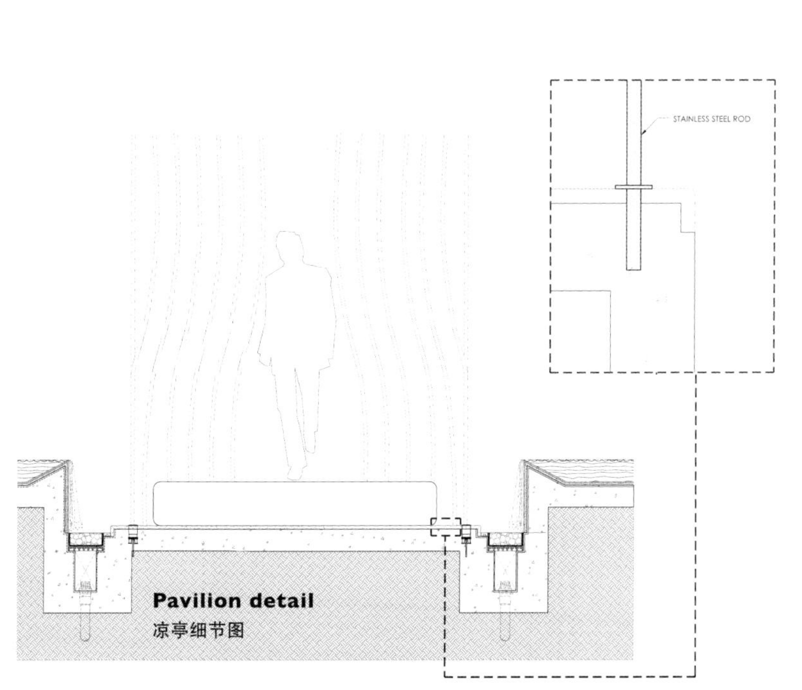

Pavilion detail
凉亭细节图

6~8. The paving in the garden
9. Lush green plants growing along the path
10. Bird eye view at night
11. Stepping stone and gravel beds

6～8. 花园的特色铺装
9. 小路两边郁郁葱葱的绿色植物
10. 夜色下的俯视图
11. 垫脚石和碎石带

10

11

SARI BY SANSIRI

新西里沙丽住宅

Location: Bangkok, Thailand
Completion: 2013
Design: Shma Company Limited
Design Director: Yossapon Boonsom
Landscape Architect: Ken Chongsuwat, Ponlawat Pootai
Photography: Mr.Pirak Anurakyawachon
Area: 1,062.41sqm

项目地点：泰国，曼谷
完成时间：2013年
设计师：Shma景观设计公司
设计总监：约撒彭・布恩索姆
景观设计师：肯・崇苏瓦，蓬拉瓦・普泰
摄影：皮拉克・阿努拉克亚瓦秋
面积：1062.41平方米

Water is necessary to all life, the landscape expresses this essence of nature through architecture and applies the design in the form of "River Delta". With a limited accommodation area for 7-floors residents, the landscape becomes most essential for its people. The project is divided into East and West buildings, leaving a gap in between to receive enough winds for all rooms. River Delta is then translated into landscape architecture and takes part in uniting the overall into one.

Characters of river are expressed richly in bold graphics of architectural environments, from finishing to spatial space. Natural elements of trees, bushes, gravels and woods are finely arranged to create an easeful environment in rigid geometrical forms. On rooftop garden, entrance ground is shifted down to connect with lower fitness floor, planters are raised to create a zigzag river walk boundary, and extended pocket space of grass lawn brings panoramic viewing experience to visitors.

The landscape is divided into 2 functioning areas on each roof garden. East building occupies by strolling area and resting area on the opposite side which both connect to lap pools that extend along the west edge. West building is for a more private couch surrounded by greenery and BBQ lawn facing east, which has a sunken walkway in between. The two sides that facing each other, are visually connected with transparent glass railing while all the rest are solid walls either are tall bush or concrete railing.

An alignment of finished materials are placed accordingly from east to west, connecting all landscape areas together. Dynamic patterns of vertical fins turn machine room into another curious volume yet camouflage soundly as part of landscape elements. The finished wall turns into an elusive light and shadow play at night, and provides shade for the pool during the day. Bush planters at the river walk slope up to wrap resting area with greenery. Trees are planted above private couch and daybed by the pool, becoming green lush scenery for both sides. At ground floor, double-height ceiling entrance finished with warmth wood texture expresses formality, in contrast to the rooftop garden.

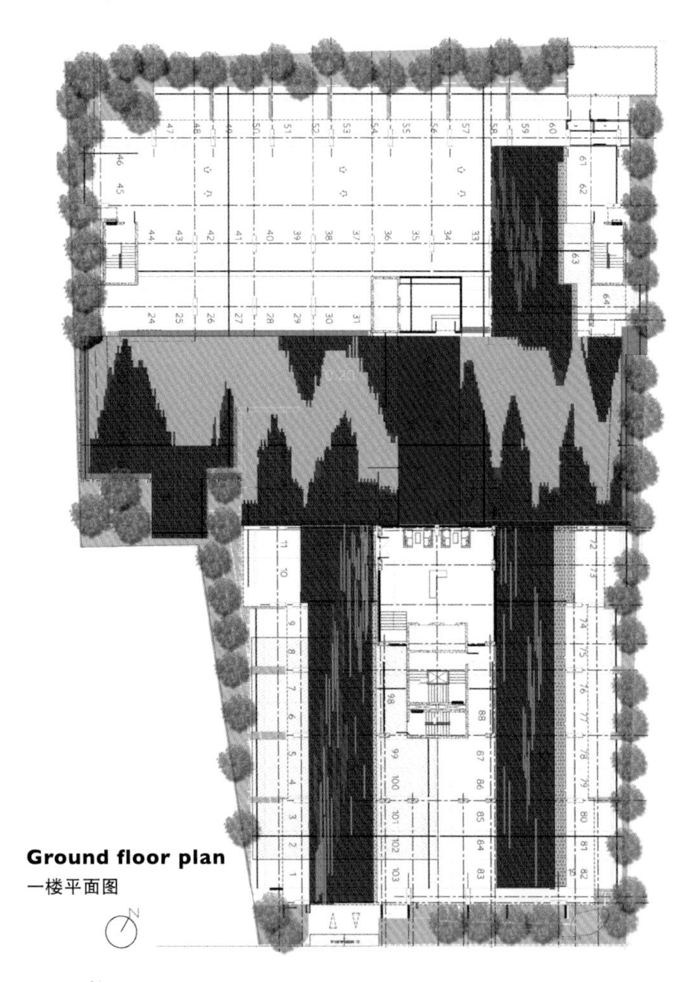

Ground floor plan
一楼平面图

Ultimately, design that accommodates greenery could drive growth of urban ecology along with the city's economic environment. It not only creates positive impact such as reducing heat which is crucial to tropical city, but also encourages outdoor activities for public health and importantly mitigates negative impacts to the urban environment.

水是生命之源，项目的景观设计通过建筑体现了这一自然基础，并且在设计中采用了三角洲的形状。在这座以 7 层高住宅楼为主的小型居住区内，景观对居民显得至关重要。项目将东西两侧的住宅楼隔开，中央的间隔空间能保证所有房间都享有自然通风。三角洲被纳入了景观设计，使整体景观融为一体。

建筑环境以鲜明的图形充分地体现了河流的特征。乔木、灌木、碎石、树林等自然元素完美地组合起来，在棱角分明的结合造型中营造出舒适安逸的环境。屋顶花园的入口地面略低，与较低的健身楼层连接起来。抬高的花池形成了曲折的河道边界，而延伸的迷你草坪则为人们带来了广阔的视野。

每个屋顶花园的景观设计都分为两个功能区。东楼是散步区和休息区，二者均与西侧边缘的游泳池相连。西楼设置着被绿树环绕的躺椅和烧烤草坪，下沉式通道将二者连接起来。两个区域面对面相对，

2

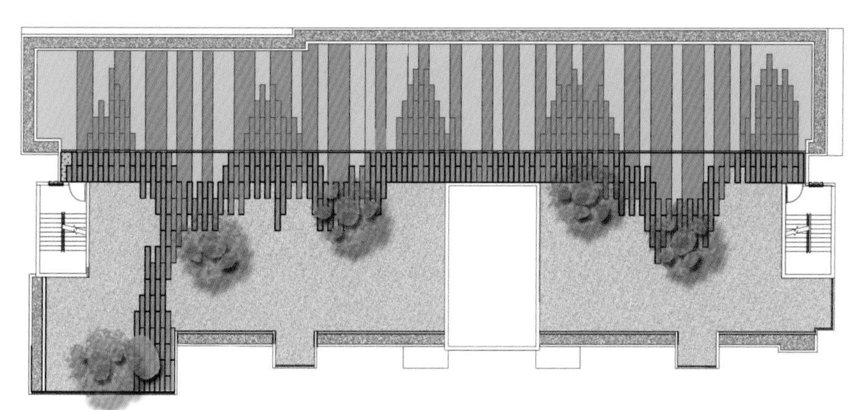

Roof floor plan
屋顶平面图

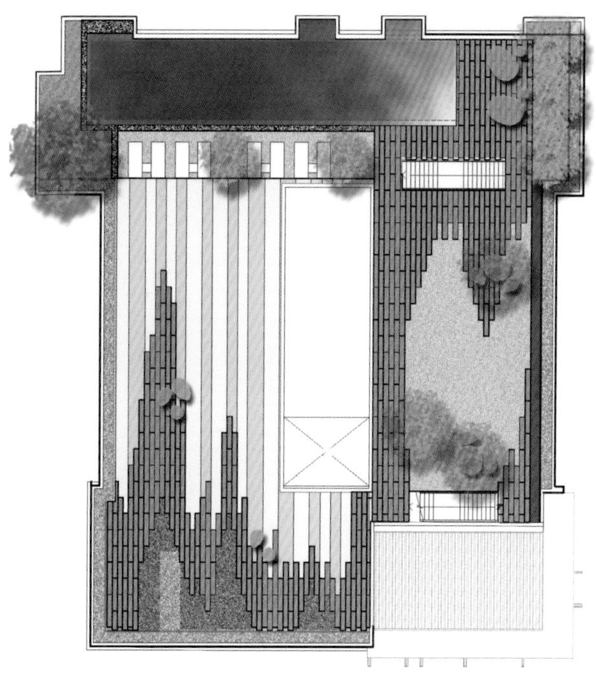

Roof floor plan
屋顶平面图

通过透明玻璃围栏在视觉上联系起来。其他方向的围栏则均为高高的灌木丛或混凝土。

一排饰面材料由东向西排列，将所有景观区域连接起来。造型动感的垂直翅片将机械室打造成一个奇妙的空间，使其伪装成景观元素的一部分。装饰墙在夜晚会呈现出迷人的光影效果，而白天则为泳池提供阴凉。河道两侧的灌木花池将休息区包围起来；长沙发和泳池边的日光浴躺椅上方种植着茂密的树木。建筑一楼的双高天花板入口采用木质纹理，十分正式，与休闲的屋顶花园形成了对比。

1. Façade view
2. Bird eye view
3. The roof garden

1. 住宅正面景致
2. 鸟瞰图
3. 屋顶花园

4. Strolling area
5. Lap pools
6. Resting area

4. 散步区
5. 西侧边缘的游泳池
6. 休息区

融入绿植的设计能够促进城市生态环境和经济环境的发展。它不仅能减少建筑的热增量(这点对热带城市尤为重要),做出积极影响;还能鼓励公众进行户外活动，改善他们的身体健康，从而减少对城市环境的消极影响。

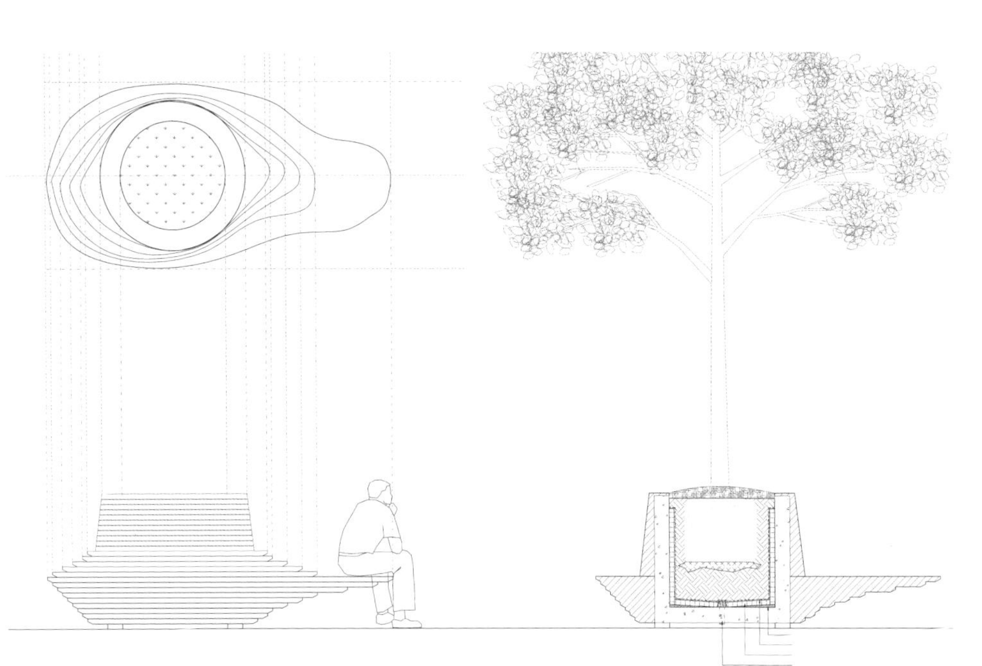

Custom seat
定制座椅

5

6

7-9. Details of the grass lawn and gravels
10. Night view of the pool
11. Night view of the strolling area

7～9. 草坪和碎石铺装细节图
10. 泳池夜景
11. 散步区夜景

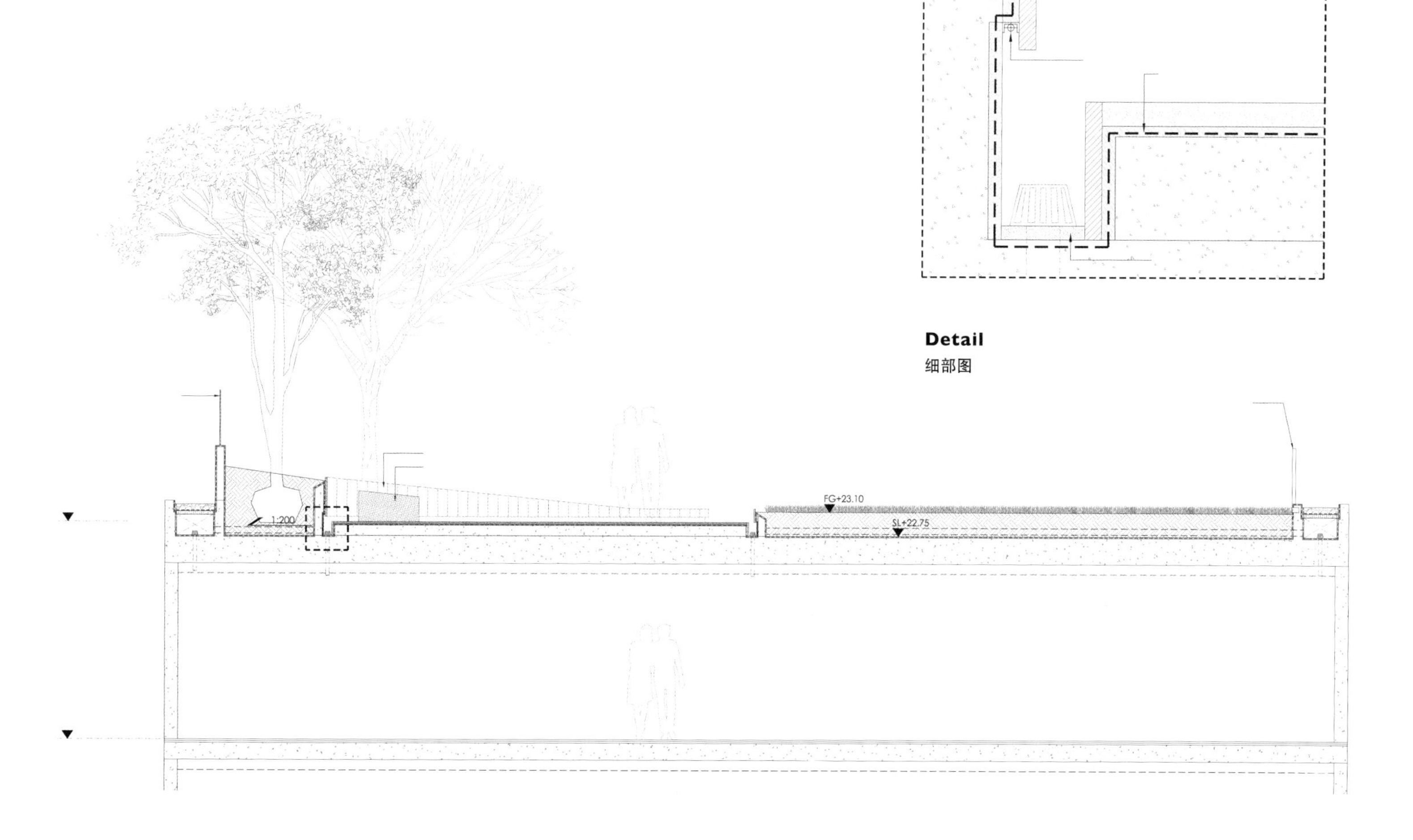

Detail
细部图

BAAN SANSUK
幸福之家

Location: Hua-Hin, Thailand
Design: TROP
Photography: Pattarapol Jormkhanngen, Pok Kobkongsanti
Area: 11,613sqm
项目地点：泰国，华欣
设计师：TROP公司
摄影：帕塔拉博尔・杰姆可汗根、
波克・柯布康桑提
面积：11,613平方米

Baan Sansuk is an exclusive residential project, located at Hua Hin, Thailand's favourite Beach. The impressive nature style condominium contains a 230-metre-long modern seascape swimming pool, which provides amusement for kids and the peaceful dreamy relaxing days for adults. Baan Sansuk is situated in the prime-location of Hua Hin beach, only a few minutes to Chatchai Market and all other conveniences.

The site is long, noodle-like with a small narrow side connected to the beach. There are 2 rows of buildings on both sides, leaving a long space in the middle of the site. Basically, most of the units, except the beach-front ones, do not have any ocean view. Instead they are facing the opposite units.

The designers' first move is to bring "the view" into the property instead. Their inspiration of "the view" comes from the location of the project. Hua Hin, in Thai, means Stone Head. The name comes from the natural stone boulders in its beach area. So the designers proposed a series of swimming pools from the lobby to the beach area, a total of 230 m long.

The Pools are divided into several types, with different functions like Reflecting Pool, Kids Pool, Transitional Pool, Jacuzzi Pool and Main Pool. At some certain area, the designers strategically place Natural Stone Boulders to mimic the Famed Local Beach. The result is a breath-taking Water Landscape, with different Water Characters from one end of the site to the other. These pools are not just for eye-pleasure only, but they also serve as the pools for everyone in the family.

幸福之家公寓是一个豪华住宅项目，位于泰国最受欢迎的海滩华欣。这个令人流连忘返的自然风公寓项目包含一个230米长的现代水景池，既为儿童提供了丰富的娱乐设施，也为成年人提供了宁静的梦想休闲空间。幸福之家公寓位于华欣海滩极好的位置，距离查柴市场和其他便利设施仅有几分钟的路程。

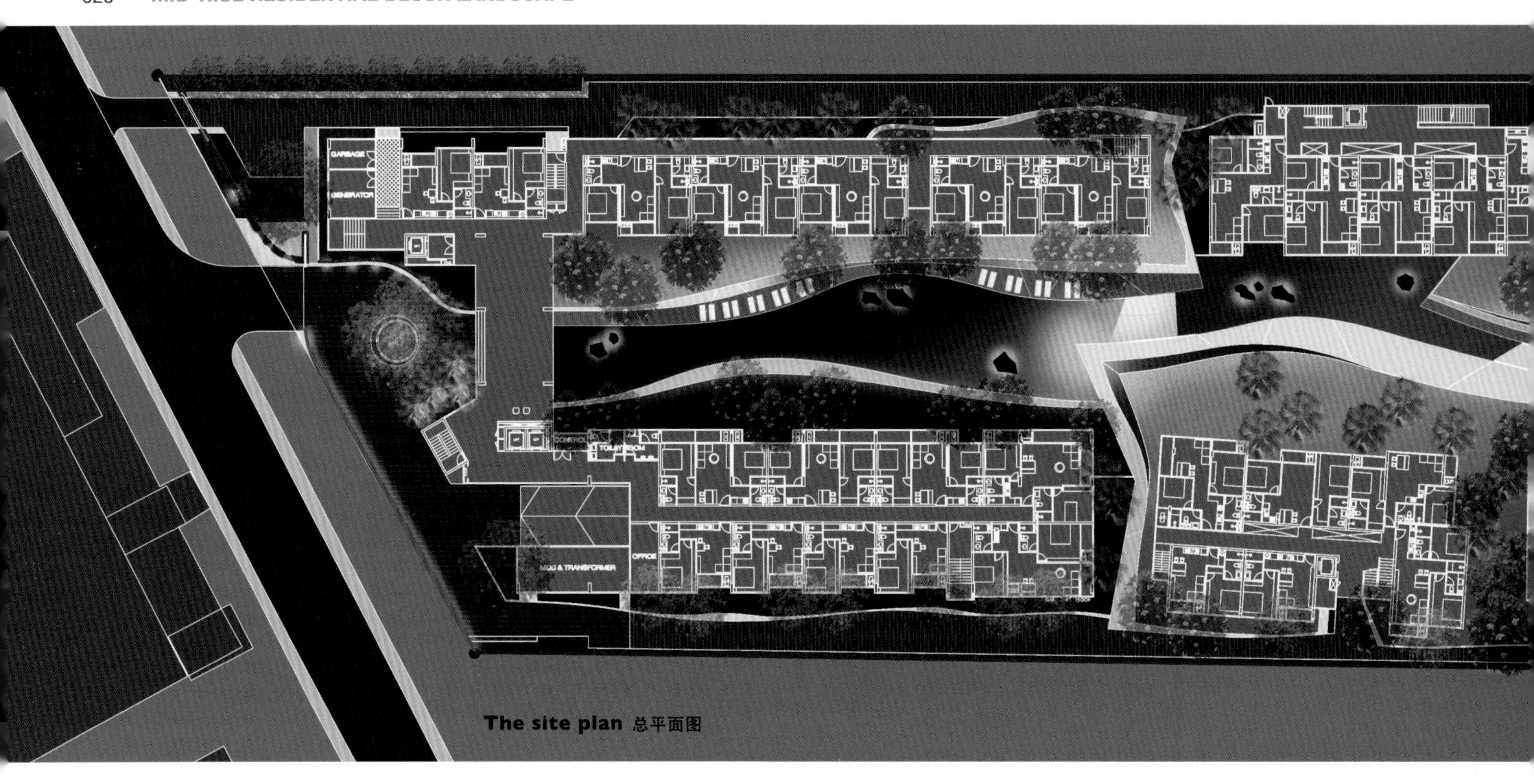

The site plan 总平面图

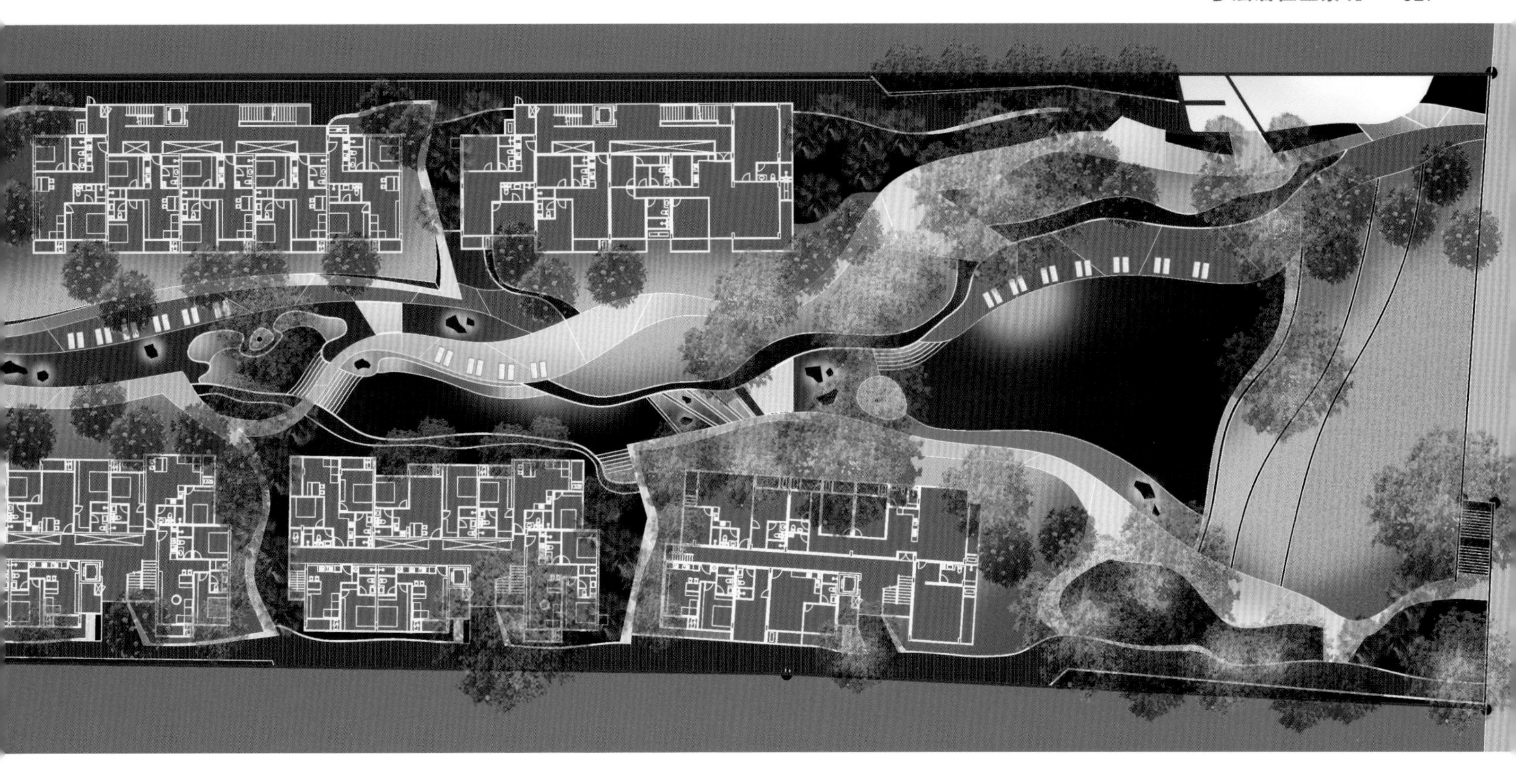

项目场地呈细长形，狭长的一边与海滩相连。项目两侧均有两排公寓楼，中间留有一块长条空间。除了正对沙滩的住宅单元之外，其他的住宅单元都不能享有海景，而是与对面的住宅相对。

设计师的第一步是将“景色”引入公寓区。他们的“景色”灵感来自于项目的所在地。“华欣”在泰语中意为“石头”，其名字来源于海滩上的巨型天然卵石。因此，设计师从公寓大厅到海滩设计了一系列大大小小的水池，总长度可达230米。

水池根据功能分为不同的类型，包括倒影池、儿童泳池、过渡池、极可意按摩浴缸和主泳池。设计师在某些区域巧妙地放置了一些巨大的天然卵石来模仿海滩。最终，设计形成了令人叹为观止的水景景观，在场地中贯穿了各种不同的水景特征。这些水池不仅具有良好的视觉效果，还能切实为家庭中的每位成员提供各种服务。

5

6

1. Landscape lighting in front of the building
2-3. The site is a long, noodle-like with a small narrow side connected to the beach
4. Water landscape
5-6. The swimming pool
7 The designers strategically place Natural Stone Boulders to mimic the Famed Local Beach at some certain area
8-9. Bird eye view of the pools

1. 公寓前面的景观照明
2、3. 项目场地呈细长形，狭长的一边与海滩相连
4. 水景景观
5、6. 游泳池
7. 设计师在某些区域巧妙地放置了一些巨大的天然卵石来模仿海滩
8、9. 水池鸟瞰图

7
8
9

BAAN SAN KRAAM
博安圣克拉姆海景公寓

Location: Cha-Am, Petchaburi, Thailand
Completion: 2013
Design: Sanitas Studio Co., Ltd.
Photography: Wison Tungthunya
Area: 23,110sqm

项目地点：泰国，碧武里，七岩镇
完成时间：2013年
设计师：萨尼塔设计工作室
摄影：威森・唐森亚
面积：23,110平方米

Sanitas Studio was commissioned in 2011 by Sansiri PLC to devise a masterplan and landscape design for Baan San Kraam, a residential project on Cha-Am beach, Petchaburi, Thailand. Only 10 per cent of its extensive area borders the beach meaning just two of the thirteen buildings have a direct sea view. The challenge for Sanitas was to provide a landscape design allowing all residents to feel equally close to the beach.

Starting with a nautical nostalgia theme, Sanitas envisaged the land as an abstract form of the ocean. It would contain different elements of islands and seascape such as jungle, villages and a floating house. Seven clusters of buildings were then created with their own unique seascape character.

The landscape design is a simple interpretation of the wave typology. Having studied its form in detail, Sanitas has developed it into three-dimensional landscape form. This includes wave seating, stepping stones, rock day beds and a tree house, while water is the landscape's key element and connects all zones together.

Residents can experience the different character of each area from their arrival at Lobby Beach and then across the water to the Jungle which is surrounded by secluded villages. Local coastal plants enhance the natural beauty of the development and its stunning oceanside concept, while existing trees have been preserved to provide welcome shaded area.

"Ocean" Concept

Starting with nautical nostalgia, Sanitas Studio envisages the land as an abstract form of the ocean. The abstract ocean would contain different characters of islands and seascape: the jungle, the villages, the floating house and the sea tide. This would create different atmosphere for each cluster of buildings. Then the designers create seven clusters of building with their own unique seascape character, from Lobby Beach, Fisherman Village, Modern Jungle, Village Tree House, Village Pool, Connecting Beach and Floating House.

1. The designer envisaged the land as an abstract form of the ocean.
2. The "ocean" concept plays an important role

1. 设计师以海洋为主题，将楼宇想象成海上的岛屿
2. 海洋主题扮演了重要的角色

Master plan
总平面图

Wave Typology

From the "ocean" concept, the water body is the key landscape element, which connects all zones together. The designers studied the natural form of the seascape and interpret it in modern way. In the overall masterplan, the form of landscape is a simplification of topography line of the ocean and uses it as the landscape language throughout the whole site.

The landscape design is a simple interpretation of the wave typology. Sanitas Studio studied the form of sea wave and developed it in the landscape form three dimensionally. It becomes part of the continuous seating, continuous steps, the retaining wall and planters.

Site Layout

At Baan San Kraam, the "ocean" concept plays an important role and connects all zones together with one unique landscape language of seascape. The water body, where all activities are met, is associated with pathway system and sunbathing deck and is partially adjacent to the balconies of some units, so people can connect to the swimming pool directly.

At Baan San Kraam, Sanitas Studio has created a series of rich details for each zone, which become its own internal view and provide a series of outdoor sitting areas, so people can enjoy resting in each unique garden. From the front entrance, the residents could arrive on the Lobby Beach with horizontal line of beachscape, across the water to the Jungle with rustic plants which is surrounded by secluded villages. There is a sunken island for Gym room, so people can exercise with a view of water cascade and have sun-bathing on the roof deck. Before arriving on the beach, there was a large swimming pool and sunbathing terraces.

At Baan San Kraam, soft planting also plays an important role; nurturing the space, differentiating characteristic of each zone, providing shade for outdoor function space and decreasing temperature, which is crucial in tropical country. The existing trees have a big impact on landscape's atmosphere, especially for each front area, where there are stunningly mature Rain tree, Flame tree and Tamarind tree.

Circulation

At Baan San Kraam, there is a vehicle access from the street with front parking under trellis structure. After arrival at the lobby, people can choose to walk through the main walkway, which connects to all buildings or choose the golf cart-service and access from the service way, which runs along the periphery. The parking is located at the entrance area and provides service for mechanical room and garbage room.

Beach pool section
海滩泳池剖面图

博安圣克拉姆海景公寓（Baan San Kraam）是泰国盛诗里房地产公司（Sansiri PLC）在碧武里府七岩小镇的海滩上新开发的楼盘。萨尼塔设计工作室（Sanitas Studio）负责整体规划和景观设计。虽然开发面积很大，但是只有10%的土地毗邻海滩，13栋公寓楼当中只有两栋直面海景。萨尼塔设计工作室面临的挑战是如何利用景观设计让所有住户都感觉离海滩很近。

设计师以海洋为主题，将整个场地想象为海洋，而楼宇就是海上的岛屿，楼宇之间漫生出“丛林”、“村庄”、“水上住宅”等海景元素。13栋建筑总共形成七个楼群，各具独特的海景特色。

景观设计是对“波浪”的一种简单诠释。在仔细研究了波浪的形态后，设计师提出了三维立体的景观形式，包括蜿蜒的座椅、踏步石、石床和树屋等。同时，水是这一景观中的关键元素，将所有空间连接在一起。

住户可以体验到园区内每个空间不同的特征。从“海滩入口”开始，然后穿过水景，来到“丛林”，“丛林”周围是隐秘的“村庄”。设计师采用了当地的滨海植物，突出了自然之美，也强化了海景的设计主题，同时保留了原有的树木，在入口处营造出一片阴凉。

“海洋”主题

设计师以海洋做为该园区的主题，将整个开发场地视为一片抽象的海洋，“海”上有海岛以及各种海景：丛林、村庄、水上住宅和海浪等，利用这些元素赋予每个“海岛”不同的特点。七个楼群各具海景特色，

3

4

5

有“入口海滩”、“渔夫村庄”、“现代丛林”、“村庄树屋”、“村庄泳池”、“连接海滩”和“水上住宅”等。

“波浪景观”

延续“海洋”的主题，所以水是这个设计的主要景观元素，是联系所有区域的关键。设计师研究了海景的天然形态，并将其转化为现代设计语言。在本案的整体规划中，景观的形态是对海浪的一种抽象简化，“波浪”是贯穿整个园区的设计语言。

设计师研究了海浪的形态，并将其表现为立体的景观形态。波浪的曲线化身为绵延的座椅、连续的踏步、围墙和花池，出现在各个位置。

整体布局

在博安圣克拉姆海景公寓的景观设计中，“海洋”主题扮演了重要角色，用一种独特的景观设计语言将各个区域联系起来。水景周围可以进行各种活动，并与四通八达的步道和日光浴平台相连。有些住户从阳台上就能直接俯瞰水景。

3. Outdoor sitting area
4. The white paving among the green plants
5. Sunbathing terrace
6. Sunken island for Gym room at night

3. 户外休闲空间
4. 绿色植物间穿插的白色铺装小路
5. 日光浴平台
6. 下沉区健身中心的夜景

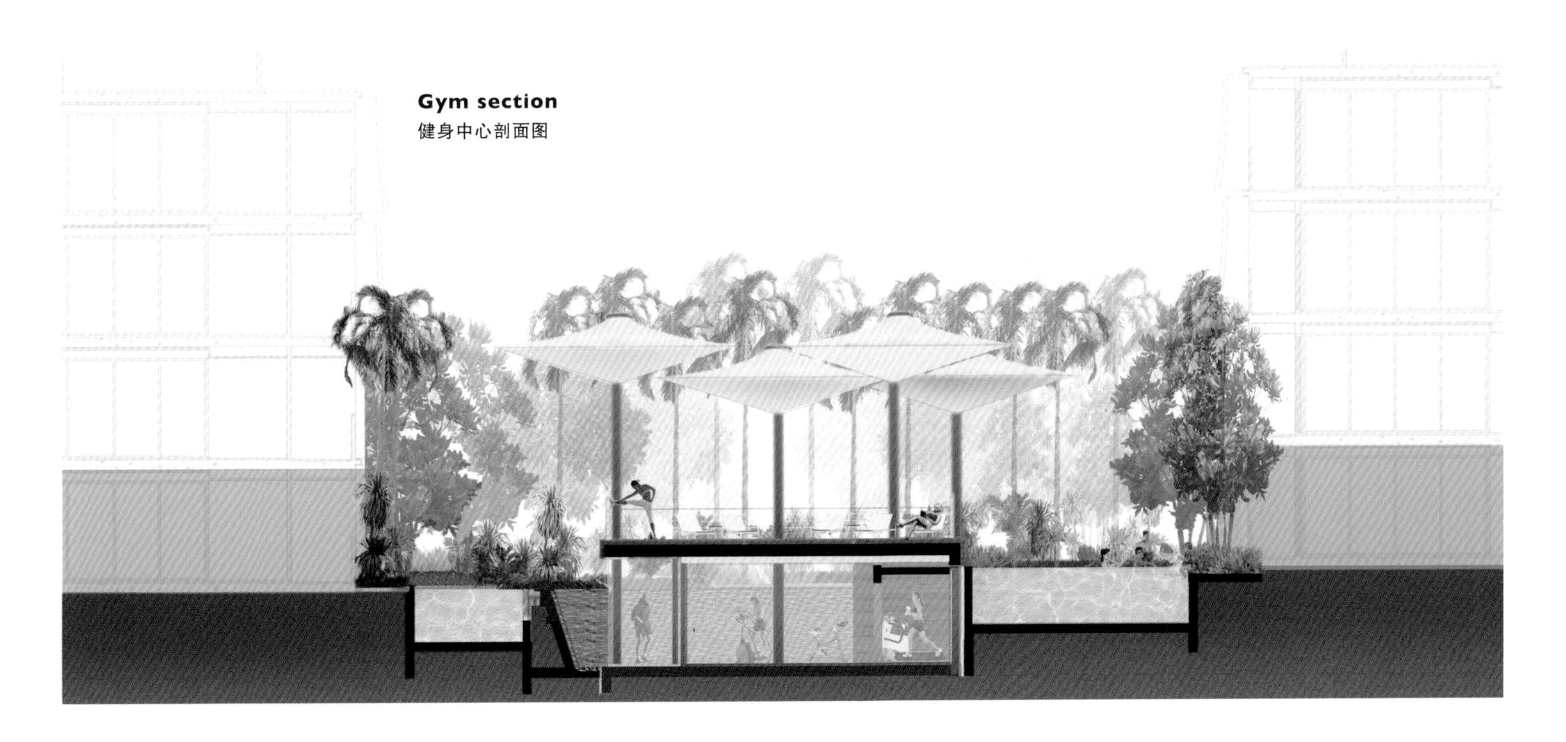

Gym section
健身中心剖面图

7. Unique seascape character
8-10. A series of rich details
11. Sunken island for Gym room in daylight

7. 独特的海景特色
8～10. 丰富的细节
11. 下沉区健身中心的日景

设计师为园区内的每个区域都设置了丰富的细节，为公寓楼打造了完美的风景，每个独特的小花园都可以成为居民的户外休闲空间。正门处是“入口海滩”，居民可以在这里享受美丽的滨海风景。穿过水景就来到“丛林”，这里有茂盛的植物。“丛林”周围是隐秘的“村庄”。下沉区的健身中心自成一个“海岛”，在这里可以一边健身一边欣赏瀑布，还可以在屋顶平台上晒日光浴。到达“海滩”之前，你还会看到一个大型泳池以及日光浴平台。

“软景观”在本案中也起到重要的作用，不仅丰富了空间内容，而且让每个区域具有不同的特点，还为户外功能空间带来阴凉，有助于降低温度——在热带国家这一点尤为重要。原有的树木极大地影响了景观氛围，尤其是每栋建筑前方的空间，其中包括雨豆树、凤凰木和罗望子树等，营造出热带风情。

交通循环

博安圣克拉姆海景公寓设置了一条机动车通道，与园区外的街道相连，入口停车场设置在藤架结构的下方。来到“入口海滩”后，你可以选择走主路（与所有建筑相连），也可以选择侧路（高尔夫手推车通道）。停车场就设置在入口处，里面还包括机械房和垃圾存放间。

11

NEO BANKSIDE
尼奥河畔

Location: London, UK
Completion: 2013
Design: Gillespies
Photography: Jason Gairn
Area: 7,700sqm
项目地点：英国，伦敦
完成时间：2013年
设计师：吉尔斯派斯
摄影：詹森·盖尔恩
面积：7,700平方米

Developed as an integral part of the residential scheme, the new landscapes at NEO Bankside provide richly-detailed green areas that balance beautifully with the contemporary apartment pavilions. Unusually in the heart of the city, the new outdoor spaces offer NEO Bankside's residents unique opportunities to engage with nature.

The landscape designs take cues from natural processes found within woodlands, and transpose them to the city. A beehive is installed, and an orchard of fruiting trees and an herb garden give residents access to produce, and add colour and fragrance to the garden areas.

As with any residential development located in the heart of a city, outdoor space was restricted, but Gillespies' designs sought to ensure that NEO Bankside's exterior spaces reached their full potential.

NEO Bankside's green spaces offer both residents and members of the public passing through a mix of the landscape typologies we find in nature. This approach creates a rich microcosm of landscapes within a constrained footprint.

Sustainability and Environmental Considerations

Gillespies' designers placed great importance on selecting the most appropriate materials for NEO Bankside's landscapes in respect to the environment, place-making and long-term performance. Gillespies specified all elements as suitable for the context, to limit impact on the environment, and where relevant, to be robust and tolerant enough for the stresses of a public environment over a long period of time.

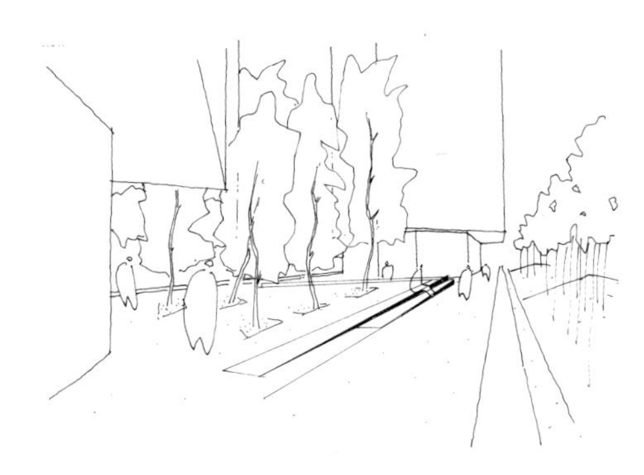

Working with planting specialists Growth Industry, Gillespies included large tracts of native plants into the design, set within groves of trees to provide a "bank" of flowers, seeds and nesting material to encourage biodiversity and a range of wildlife to the space.

I

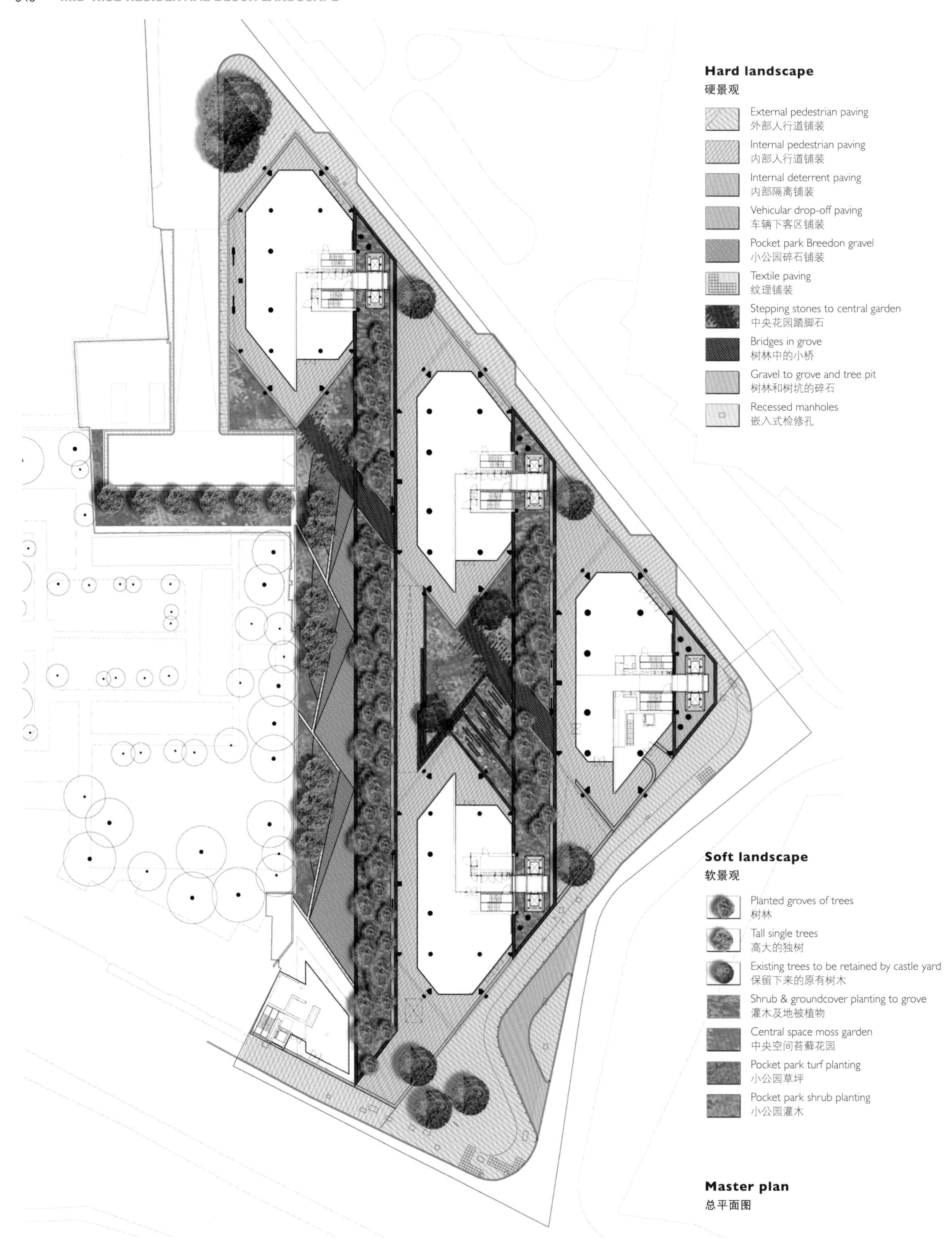

Master plan
总平面图

Rainwater Harvesting for Irrigation

Working with Hoare Lea engineers, the scheme evolved to ensure capacity for rainwater harvesting was a central tenet of the basement design and construction. Water retention boards (reservoirs) were laid over the structural slab – this technology provides a reserve of water to maintain soil saturation and consequently limits the amount of irrigation water required. This reserve of water supplements the planting irrigation system, and limits the demands on mains water use.

Planting and Biodiversity

The planting concepts and final details are central to the overall landscape design for NEO Bankside. The planting softens the built environment, humanises the space and mitigates the local microclimate to create comfortable, welcoming spaces. It also provides a seasonal sense of time and place to enrich urban life. Most of the plants used at NEO Bankside are native in origin and are carefully suited to the microclimate of the site. BREEAM guidelines and biodiversity were major drivers for the selection of appropriate plant species.

2

3

1. The new landscapes provide richly-detailed green areas that balance beautifully with the contemporary apartment pavilions
2-3. The green spaces offer both residents and members of the public passing through a mix of the landscape typologies we find in nature

1. 新景观以精致而丰富的绿地与现代公寓楼交相辉映
2、3. 绿化空间为居民和过往的行人提供了与自然环境类似的丰富景观类型

4. Bird eye view of the overall landscape
5. The detail of the paving
6. The side view of the sculpture
7. The detail of the flower bed
8. The planting softens the built environment
9-11. A "bank" of flowers

4. 整个景观区的鸟瞰图
5. 铺装细节图
6. 雕塑的侧面视角
7. 花池细节图
8. 植物柔化了建筑环境
9～11. 设计中引入了大量的植物

8

9

10

11

作为住宅项目的重要组成部分，尼奥河畔的新景观以精致而丰富的绿地与现代公寓楼交相辉映。新的户外空间为尼奥河畔的居民提供了与大自然亲密接触的机会，这在市中心是罕见的。

景观设计从林地的自然过程中获得了灵感并将其带到了城市之中。项目安装了封箱，添加了一个小果园和一个香草园，让居民们可以享受生产的快乐，同时也为花园区增添了色彩和芳香。

与其他位于市中心的住宅项目相同，该项目的户外空间也十分有限。但是设计师最大限度地利用了尼奥河畔的户外空间，充分发挥了它们的潜力。尼奥河畔的绿化空间为居民和过往的行人提供了与自然环境类似的丰富景观类型，在有限的空间内打造了丰富的微型景观世界。

可持续特征和环保措施

设计师在景观材料的选择上花费了大量的经历，综合考量了环境保护、场所营造和长期表现等因素。他们详细列出了适合该园区环境的景观元素，既考虑了它们对环境的影响，又挑选了足以长期抵抗公共环境压力的材料。

设计师与栽培顾问成长工业公司共同合作，在设计中引入了大量本土植物，在树林中提供了花岸、种子和筑巢材料，促进了生物多样性，也吸引了更多的野生动物前来定居。

收集雨水用于灌溉

景观设计师与霍尔·利工程公司合作，将雨水收集作为地下设计和建造的中心环节。水分保持板被覆盖在结构板之上，这一技术利用水储备保证了土壤处于饱和状态，从而限制了灌溉水的需求。这些水储备能补充植物灌溉系统，减少总用水量。

植物栽种与生物多样性

种植理念与细节完善是整个景观设计的中心。植物柔化了建筑环境，让空间人性化，并且调节了本地微气候，营造出更舒适、更友好的空间。此外，植物还体现了季节的变化感，丰富了城市生活。设计师所选用的植物大多来自于本地，与项目场地的微环境十分合适。英国建筑研究所环境评估法（BREEAM）指导原则和生物多样性是选择合适植物种类的主要驱动力。

BLOCK 32 AT RINO
北河32号街区

Location: Denver, USA
Completion: 2013
Design: studioINSITE
Photography: studioINSITE
Area: 16,187sqm
Awards: 2013 Denver Mayor's Design Award, 2013 Denver Multifamily Project of the Year, awarded by the Rocky Mountain Real Estate Expo

项目地点：美国，丹佛
完成时间：2013年
设计师：INSITE工作室
摄影：INSITE工作室
面积：16,187平方米
获奖：2013丹佛市市长设计奖、2013落基山脉房产展览会丹佛多户住宅项目年度大奖

The Block 32 at RiNo project offers two hundred and five multi-family rental housing units within the burgeoning RiverNorth (RiNo) District just blocks north of downtown Denver, Colorado. The space is designed to celebrate the unique urban character of the up-and-coming RiNo Art District, with its high concentration of creative businesses and array of studio spaces in an industrial environment. studioINSITE provided site design and landscape architecture to the client for the four acre site.

The Block 32 development has been designed with the flexibility for potential ground-level commercial space, allowing for a live-work community. Block 32 at RiNo is constructed around a shared courtyard that offers a space for continued community interaction. With space for both active and passive use around planted areas, a swimming pool and spa, the Block 32 courtyard remains an active amenity zone. Residents enjoy the outdoor patios, completed with outdoor televisions, speakers, barbeques, foosball, ping-pong and even a bocce ball court.

The building's bold colours of red and yellow are echoed in the playful red site furniture. Plantings at the site include native grasses, bamboo, yucca and bright blooming annuals. The selected trees include Thornless Cockspur Hawthorn, Autumn Brilliance Serviceberry, Japanese Tree Lilac, and Chanticleer Pear. The design uses industrial materials in unusual ways to solve a variety of functional issues while reinforcing the "artistic" characteristics of the project. Playful shapes can be found throughout; the swimming pool and topographical courtyard landscape forms take on organic and unexpected shapes.

Sustainable features are a prominent part of the project. The complex is designed to meet Enterprise Green Community standards. Highly visible along Brighton Boulevard, a solar panel trellis is one of the focal points of the project. It showcases the use of renewable energy while also providing shade. Due to urban storm

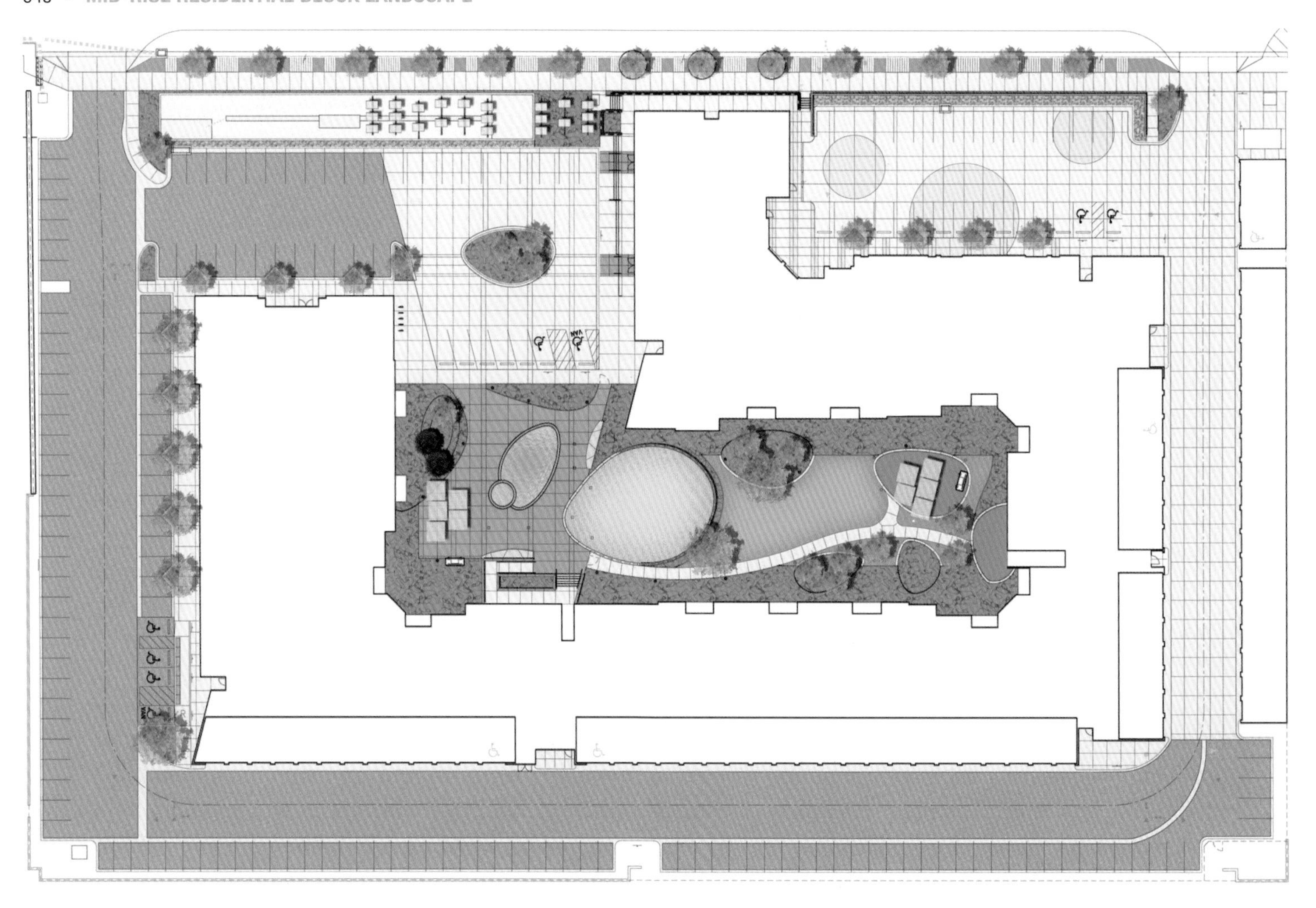

Master plan
总平面图

water drainage concerns, studioINSITE incorporated a state-of-the-art detention facility into the landscape architecture for the property, using an imaginative wall to integrate it with the character of the neighbourhood.

Helping to transform the face of the RiverNorth district, Block 32 at RiNohas earned recognition as part of an "area of change" in the Denver Metropolitan Area. The project has won two awards: the 2013 Denver Mayor's Design Award, and the 2013 Denver Multifamily Project of the Year, awarded by the Rocky Mountain Real Estate Expo.

北河32号街区项目为205套多户出租单元，位于美国丹佛市中心以北仅几个街区路程的北河区。北河区艺术区集中了大量创意产业和工业工作室，建筑空间体现了它欣欣向荣的城市特色。INSITE工作室为委托人提供了场地设计以景观设计服务。

32街区项目的设计具有高度灵活性，预留有地面商业空间，旨在建成一个办公兼生活社区。项目以共享庭院为中心向外拓展，促进了社区居民的互动。项目配有大量各式绿化空间、游泳池和按摩泳池，其中32街区庭院是最活跃的休闲娱乐场所。天井配有露天电视机、音箱、烤肉架、桌上足球、乒乓球台乃至室外地掷球场，居民可以尽情享受。

建筑红黄两色明艳的色彩在红色的街景家具中得到了体现。绿化种植包括天然草、竹子、丝兰及亮丽的一年生开花植物。项目栽种的树木包括无刺鸡距山楂树、秋彩唐棣、日本丁香树和雄鸡梨树。设计采用非常规的方式利用工业材料解决了一系列实用问题，同时也突出了项目的“艺术感”。到处都是有趣的造型：游泳池和庭院景观都呈现出意想不到的有机造型。

可持续特征是项目很重要的一部分，项目以企业绿色社区标准为设计目标。从布莱顿林荫大道上可以清晰地看到格栅式的太阳能电池板，它们既展示了可再生能源的利用，又能提供阴凉。考虑到城市雨水排水，INSITE工作室在景观设计中引入了先进的雨水滞留设施，并利用一面想象中的墙壁将它与小区建设结合起来。

32街区项目帮助改变了北河区的外观，是丹佛大都市区“区域改造”的一部分。项目获得了两项大奖，即2013丹佛市长设计奖和2013落基山脉房产展览会丹佛多户住宅项目年度大奖。

1. The shared courtyard landscape
2. Elliptical green lawn
3. The Block 32 courtyard remains an active amenity zone

1. 共享庭院景观
2. 椭圆形的绿色草坪
3. 32街区庭院是最活跃的休闲娱乐场所

NO
DIVING

4. The building's bold colours of red and yellow are echoed in the playful red site furniture
5-6. Details of the swimming pool
7-9. Plantings at the site include native grasses, bamboo, yucca and bright blooming annuals

4. 建筑红黄两色明艳的色彩在红色的街景家具中得到了体现
5、6. 泳池的细节图
7～9. 绿化种植包括天然草、竹子、丝兰及亮丽的一年生开花植物

ROLFSBUKTA
罗尔夫斯湾

Location: Oslo, Norway
Design: Bjørbekk & Lindheim AS
Architect: ARCASA architects, Norway
Photography: Bjørbekk & Lindheim AS
Area: 66,000sqm

项目地点：挪威，奥斯陆
设计师：贝吉比克&林登景观事务所
建筑设计师：ARCASA挪威建筑事务所
摄影：贝吉比克&林登景观事务所
面积：66,000平方米

Rolfsbukta is a bay at the north-east end of Fornebu. It is one of the few points where residential areas on the site of the old airport will have direct contact with the sea.

Very early in the project a decision was reached to reinforce the relationship of the bay with the sea and consequently a canal was extended into the bay. To bring the water as close as possible to the public, the landscape designer planned the inner 2/3 of the canal as a freshwater canal and the last 1/3 as a deep saltwater canal, connected by a waterfall between the two levels.

The residential area, Pollen, surrounding the inner part of the fresh water canal, is already completed. A large pool, a pond with stepping stones, a wooden pier and a fountain frame two sides of the Pollen buildings. The water surface is enclosed by a poured concrete ramp that slopes down on one side, and on the other sides stairs lead down to a 20 centimetre deep pool. The canal is surrounded by formal beds of ornamental grasses and willow trees framed by non-corrosive steel, the same material as is used in a custom-made barbecue. There are several seats surrounding the pool made of poured concrete with wooden covering recessed into the concrete. There is also a large platform of trees planted in gravel together with long tables and benches.

The seating, the "island" and the bridge are lit from below creating the impression that they are floating as the dark falls. More lighting is directed up towards the trees from the planter boxes, throughout the platforms, on the "island" and at the base of the spray nozzles of the water fountain. Various species of willows and cherry trees are planted close to the canal together with the ornamental grasses Chinese Silver Grass, Purple Moor Grass and Silver Feather Grass.

Further out at Rolfsbukta the second part of Phase 1, Tangen and Marina, have been completed. The pier motif with promenade is an essential part of this housing complex that is made up of 6 blocks on a north-west facing slope down towards

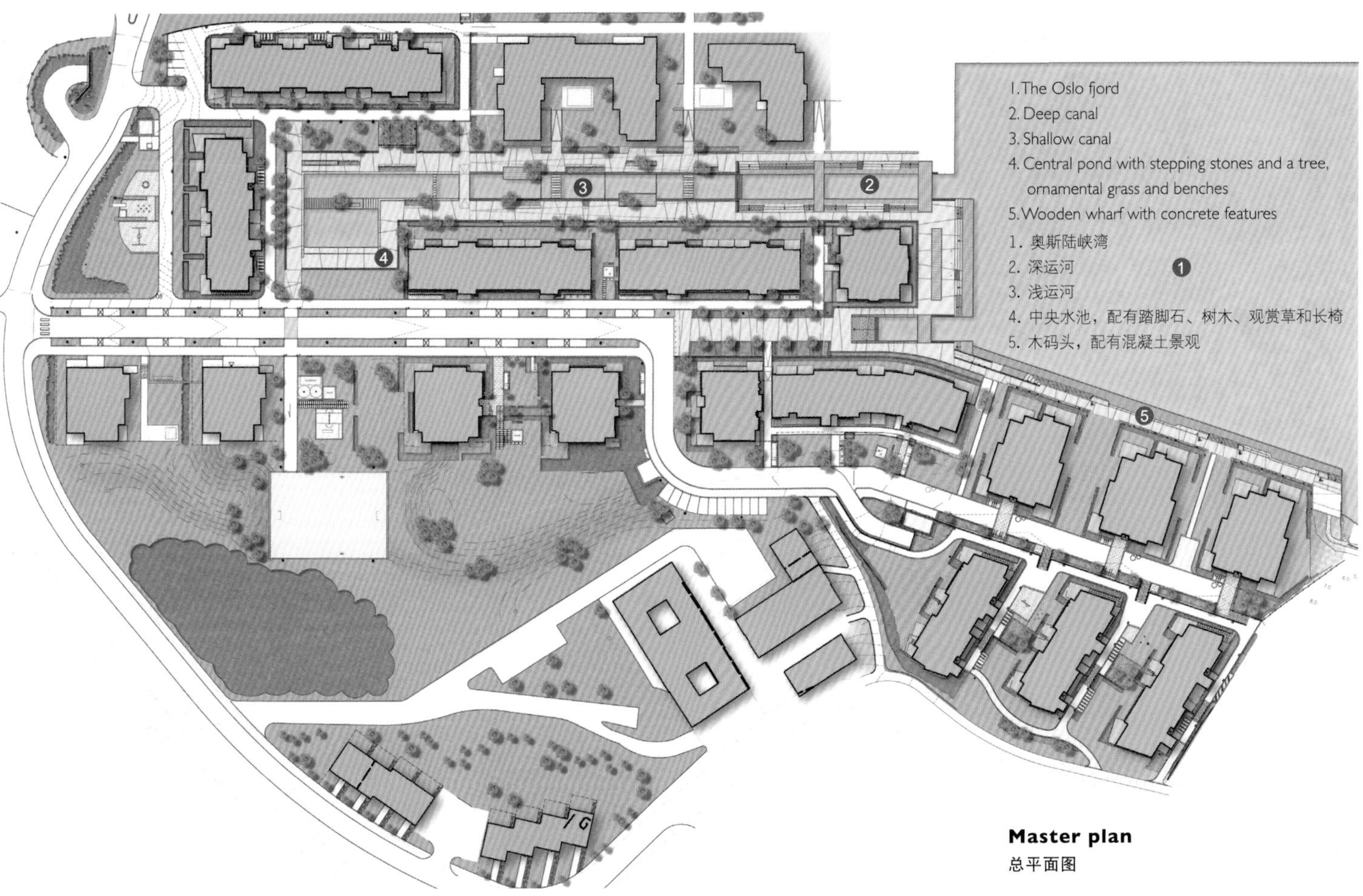

Master plan
总平面图

the sea. You can moor boats and walk out along the bay to the outmost tip of the bay. This sunny west-facing waterfront is designed for recreation. Poured concrete embankments provide steps and seating.

Ground-floor front gardens have a garden raised by about 80 cm and planted with a screen of vegetation to protect privacy. Within the Marina there is a car-free, green, spacious residential street with stairs to the upper residential area, Tangen. Custom made granite stones engraved with the addresses on each level offer a place to sit down and also demarcate each entrance.

Paths between the blocks provide good access to the sea also for those who live in the back rows. Because views of the sea and sun are meaningful, small trees, such as magnolia, Japanese Judas-trees, cherry and rowan trees have been chosen. Alpine currant bushes, beech, purple osier willows, virginia creeper, lilac, waxberries, black chokeberries, juneberries and common ivy are other plant types to be found in the area.

罗尔夫斯湾位于福内布东北端，是旧机场住宅区为数不多的能直接与大海相连的地点之一。

在项目的早期，规划者就决定加强海湾与大海的联系并将一条运河延伸到了海湾。为了让水更贴近公众，景观设计师将运河靠内的2/3设计成淡水运河，剩下的1/3则设计成深海水运河，二者由一个跌水景观连接起来。

波伦住宅区环绕着淡水运河的内部，已经完工。住宅楼的两侧分别是大水池、带有踏脚石的池塘、木码头和喷泉。水面的一侧由浇筑的混凝土坡环绕，另两侧则是通往20厘米深水池的台阶。运河被种有绿草和柳树的装饰性花池所环绕，花池所采用的耐腐蚀钢与定制烧烤

1. The residential area surrounds the inner part of the fresh water
2. A car-free spacious residential street
3. Stepping stones
4. Water fountain
5. The pond-area with stepping stones, a wooden "island", a water fountain, surrounded by formal beds of ornamental grasses, willow trees and concrete benches

1. 住区中心的淡水池
2. 宽敞的住宅区步行街道
3. 步行石桥
4. 喷泉
5. 水池里铺有步行石桥，一个木质“小岛”，一个喷泉，周围是装饰性的草坪，柳树和混凝土长凳

2

3
4
5

6

7

架的制作材料是相同的。水池旁环绕着一些混凝土制成的座椅，座椅上方覆盖着木板条。此外，小区里还有一个种植着树木、铺着碎石的大平台，旁边配有长桌和长椅。

座椅、“水中小岛”和木板桥都采用下方照明，在夜晚营造出一种漂浮于水面的感觉。更多的灯光从花池直接打在树木上，穿过平台，到达“水中小岛”和喷泉底部的喷头上。各种品种的柳树和樱桃树与中国荻草、紫酸沼草和银羽草等装饰性草种共同紧密环绕着运河。

更靠近罗尔夫斯湾的一期工程第二部分——唐恩住宅区和码头区也已经完工。码头主题和散步路是该住宅区的重要组成部分。该住宅区由六座住宅楼组成，位于朝向西北方海岸的斜坡上。人们可以停靠小船，沿着海湾走向海湾的最远端。这个阳光明媚的西向滨海空间是一个休闲区。混凝土堤坝上设置着台阶和座椅。

6. The sunny north-west facing bench is designed for recreation
7. Benches of poured concrete with wooden covering surrounds the pond-area
8. Extensive use of corten steel in the details
9. Ornamental grass
10. Silver feather grass, stipa calamagrostis

6. 向阳的木板座椅可供人休闲
7. 水边长凳中间注入了混凝土，外面覆盖有木板
8. 元素细节采用了大量的耐候钢材料
9. 观赏草
10. 针茅小叶

MODE 61
时尚61号

Location: Bangkok, Thailand
Design: Shma Company Limited
Photography: Pirak Anurakyawachon
Area: 3,200sqm

地点：泰国，曼谷
设计师：Shma景观设计公司
摄影：匹拉克·阿努拉亚瓦宗
（Pirak Anurakyawachon）
面积：3,200平方米

In response to dense urban surroundings of Bangkok, the main landscape area is conceived as an internal courtyard to create an inverted tranquility with all surrounding units looking into garden spaces. The landscape space is split in multi-level to achieve privacy between garden space and the units within limited space and also to create a dynamic view from the balcony looking down. This is resulted in three-dimensional plays of water and greenery; from reflective pond to water wall, from green roof to green wall.

At entrance, a relationship with public was taken into consideration. Rather than having a solid surface at the front boundary wall, a row of green hedge at 1.8-meter-high coupled with another layer of bamboo aligned behind it standing up to 6 meters high is incorporated to form a boundary. A natural environment from these layers of green is given not only to act as a soft screen for residential units facing to the main street but also to fulfill a public realm by adding greenery to this relatively dry neighborhood.

By entering to drop-off area, one faces with warm feature wooden wall protecting private area from exposure to chaotic pedestrians. The threshold lead people to enter from the side and while they are approaching lobby, a delicate white sculpture stands before their eyes with layers of green and water body as a background. The alfresco lobby is designed to locate under green roof which helps to absorb tropical heat and to be surrounded by cascade water which provides passive cooling without air-conditioner to reduce electrical power consumptions.

From lobby to the signature swimming pool area, wooden steps accompanied with overflow cascade by the side lead people to escalate up infusing into another space. The level of pool deck is carefully determined to elevate slightly from the lobby and to drop by 2 meters from the units on the first floor to give a sense of privacy for each areas. Changing of levels is crafted with water wall and green wall creating a tranquil ambience as a backdrop for the pool area. Willow is selected as a feature tree to envelop the pool space for its feather light leaf draping down

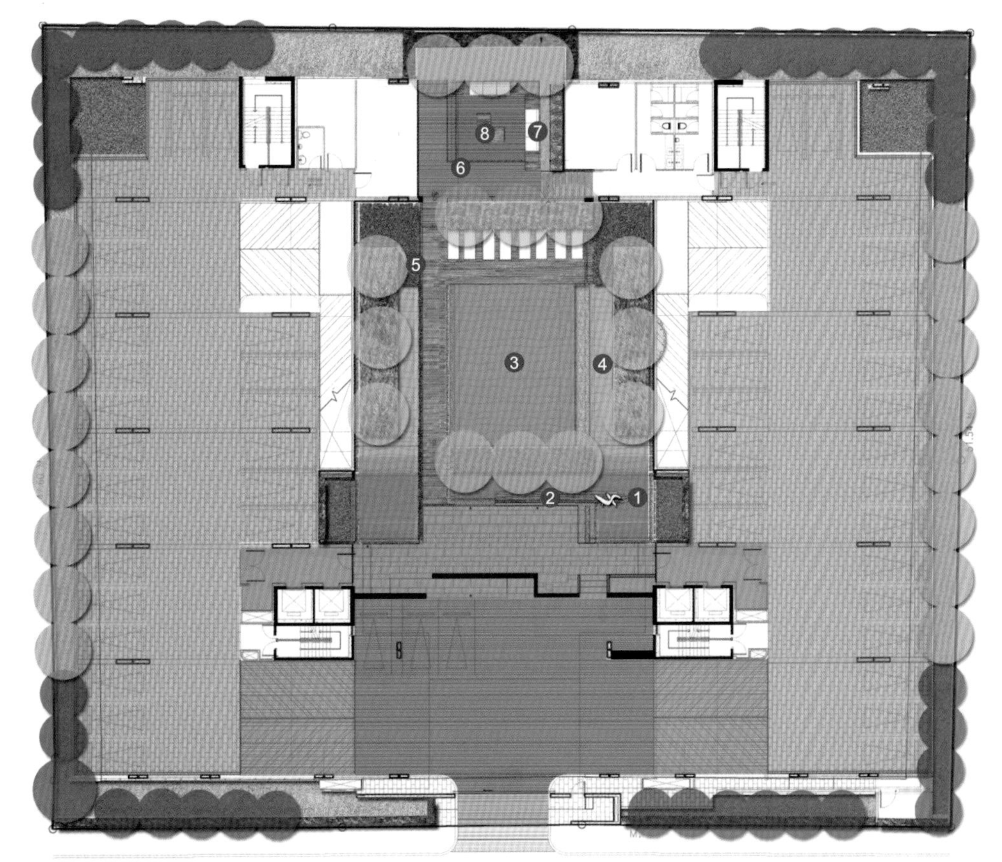

1. Overview
2. Sunken lounge
3. Sunken lounge with built-in sofa

1. 全景
2. 下沉休闲区
3. 下沉休闲区内设有固定沙发

1. Sculpture
2. Timber bench
3. Swimming pool
4. Water wall
5. Green wall
6. Sunken lounge
7. Sofa
8. Timber table

1. 雕塑
2. 木质长椅
3. 泳池
4. 水景墙
5. 绿墙
6. 下沉休闲区
7. 沙发
8. 木质桌子

Master plan
总平面图

2

forming a natural blind for the privacy of the room and swimming pool.

Consequently, moving through the pool deck to the sunken lounge where built-in reclining sofa enclaves amongst plantations. The space is designed to serve family lounging or even small party/gathering in the garden setting next to the reading room. When needed, sliding glass door of the reading room can be fully open connecting to the sunken lounge space forming a larger function spaces. The stone paving material extends horizontally to the adjacent reading room to emphasize a continuum of outdoor and indoor spaces.

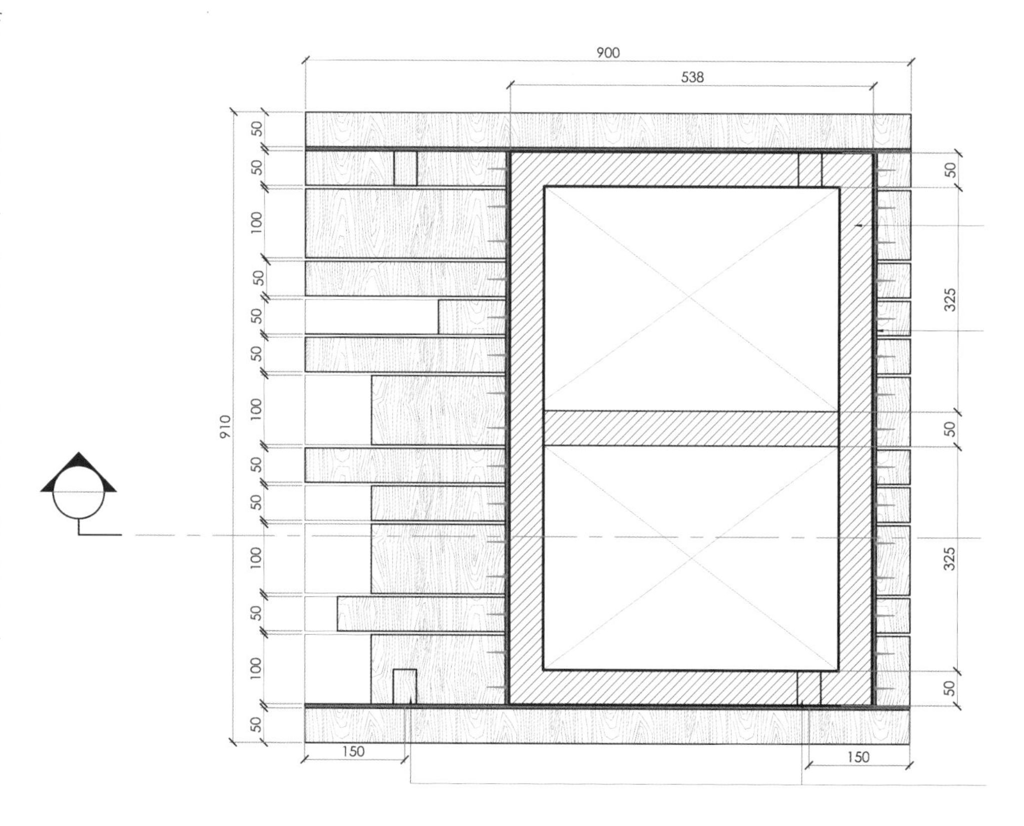

Timber table sectional plan
木质桌子部分平面图

3

The landscape of Mode 61 is created as a private green nest instilling a contemplative living environment through understated space composition and meticulous detailing. Altogether, surfaces finished in a well-defined natural stone and enriched with intricate detailing of wood work from the entrance wall to outdoor lounges, bring about an urban yet rustic outdoor dwelling. Ultimately, all meticulous detailing is curated modernly yet bio-mimicking contemplative composition.

时尚61号（Mode 61）是盖颂房地产公司（Gaysorn Property）在曼谷新开发的楼盘，园区景观设计由泰国Shma景观设计公司（Shma Company Limited）操刀。曼谷的城市环境非常拥挤，因此，设计师将时尚61号园区内的主要景观空间打造成一个中央小花园，周围的住户都可以欣赏花园内宁静的风景。设计师进一步将景观空间划分成多个层次，既保证了小空间的私密性，又丰富了景观的视觉体验，住户从阳台上俯瞰，能够欣赏到多样化的景观空间。上述目标通过水景与植被的三维立体应用而得以实现，从倒影池到瀑布，从绿色屋顶到绿墙，不一而足。

在园区入口处，设计师尤其注重考虑了小区与外界的关系，因此没有设封闭式围墙，取而代之的是1.8米高的树篱，后面搭配一行6米高的竹子，共同构成一道层次分明的绿色屏障。这道屏障营造出园区内大自然般的环境，它不仅是保护着朝向主街的单元的一道“软屏障”，而且也美化了城市公共空间，为这一相对来说比较乏味的街区平添了一抹绿意。

走进时尚61号园区，你会立刻看到暖色调的特色木板墙，保护着园区内的私密空间免于暴露在外面嘈杂的环境中。一道门槛指引着你从侧面进来，随着你走近露天大厅，一件精致的白色雕塑品便会映入你的眼帘，雕塑的背景是层次分明的绿色植被和水景。露天大厅在绿色屋顶的保护伞下，屋顶上的植被能吸收热带的炎热阳光。大厅周围还有瀑布，具有降低空气温度的作用，不必用空调，减少了电力消耗。

从大厅来到特色泳池区，会经过一段木质台阶，旁边伴随着哗哗的流水。拾阶而上，不知不觉就来到另一个空间。设计师特别将泳池平台的高度设置得比大厅略高一些（在一楼住户下方2米处），这样一来，就为每个空间都营造

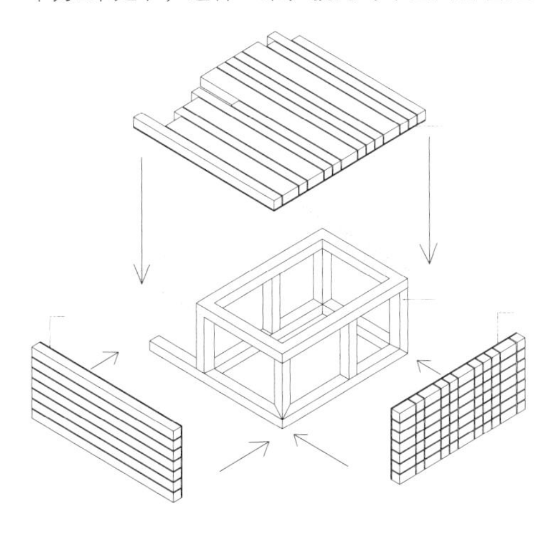

出良好的私密性。地面高差处设置了水幕墙和绿墙，为泳池区打造出静谧的背景环境。设计师选择了柳树作为主要的景观植物，环绕着泳池区种植了一圈柳树，用意是用垂柳轻盈的枝叶营造出一道自然屏风，确保泳池的私密性。

接下来，从泳池来到下沉的休闲区，休闲区内设置了嵌入式的斜背沙发，背景是茂盛的植物。这个空间的设计以家庭休闲娱乐为主，阅览室旁边的小花园里也可以举行小型的聚会。需要的话，阅览室的滑动玻璃门可以完全拉开，将阅览室和下沉休闲区连成一体，形成一个宽敞的大空间。地面的石材铺装一直延伸到阅览室里面，相同的地面材料凸显了室内外空间的延续。

时尚61号园区的景观设计通过朴素的空间布局和精致的细节处理，营造出一个私密的绿色花园，一个安静祥和的宜居环境。精致的天然石材搭配温暖的木材，从入口墙面到户外休闲区，既有现代都市的简洁，又有乡村环境的淳朴。精致的细节既体现出现代感，又模仿了大自然的有机形态。

4. The plantings
5. Paving material extends from the sunken lounge to the adjacent reading room
6. Delicate white sculpture and timber bench

4. 植栽
5. 相同的铺装材料从下沉休闲区延伸到旁边的阅览室
6. 白色雕塑和木质长椅

Feature Plants

Large Tree
- Salix babylonica Linn
- Indian Oak
- Bamboo
- Indian Cork Tree
- Mahogany

Small Plants
- Iris
- Ophiopogon japonicus (Kyoto Dwarf)
- Dwarf Umbrella Tree
- Spike Moss Family
- Alocasia Indica (Elephant Ear)
- Bromeliad
- Succulent

特色植物

大型树木：
- 垂柳
- 印度橡树
- 竹子
- 印度栓皮栎
- 桃花芯木

小型植物：
- 鸢尾
- 矮生麦冬
- 矮生鹅掌柴
- 卷柏
- 海芋（“象耳”）
- 凤梨
- 肉质植物

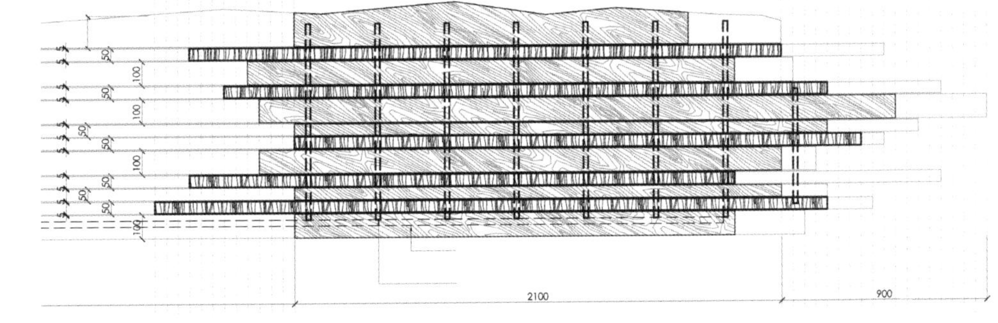

Timber bench plan
木质长椅平面图

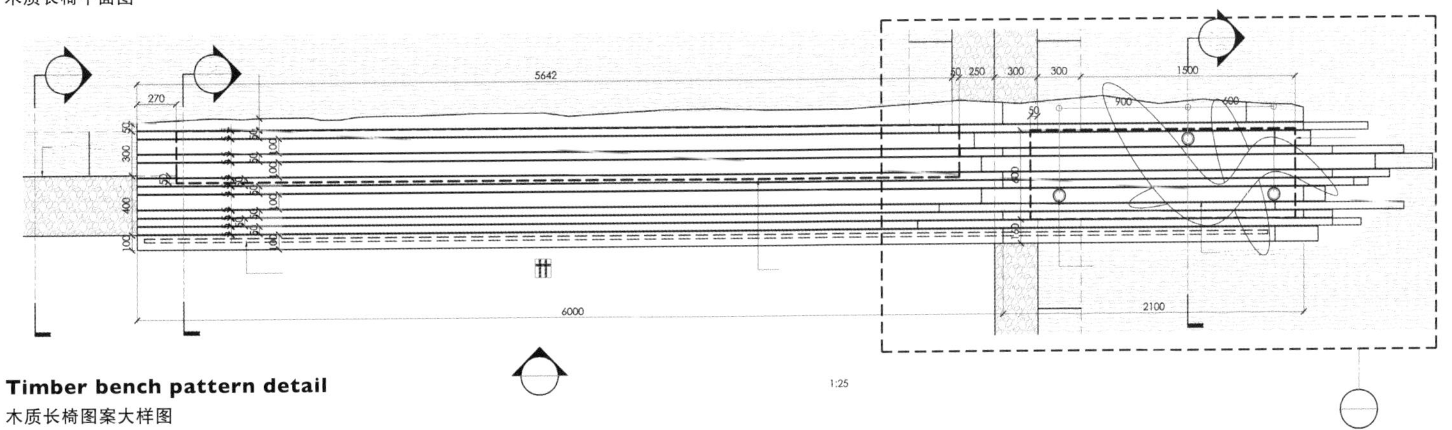

Timber bench pattern detail
木质长椅图案大样图

RENOVATION OF OUTDOOR AREAS BISPEHAVEN HOUSING ESTATE

比斯普海文住宅区户外景观翻修

Location: Aarhus, Denmark
Design: Vibeke Rønnow Landskabsarkitekter, C. F. Møller Architects
Photography: Helene Hoyer Mikkelsen
Area: 170,000sqm
项目地点：丹麦，奥尔胡斯
设计师：维贝克·罗诺景观建筑事务所、C·F·穆勒建筑事务所
摄影：海琳·霍耶·米克尔森
面积：170,000平方米

The outdoor areas around the Bispehaven social housing estate, built in 1970, were characterised by extensive surfaced areas, high concrete walls and dense planting, producing a dull and insecure environment.

Despite the attractive flats and the near-town location, the area had acquired the reputation of being something of a ghetto, with much vandalism and difficulty in attracting tenants. The aim of the total renovation project has been to give the area a new identity and quality, in particular by creating new meeting-places for the residents.

The outdoor areas have been given a sense of openness and simplicity by replacing the dense planting with lawns and flower beds. The distinctive slopes surrounding the development have been cleared of scrub and have become open grassy surfaces with sculptural rows of birch trees. The former high concrete walls have been sawn of near ground level, to become low balustrades.

The new paving uses a restrained palette of black and white patterns, to match the renovated facades of the housing scheme, and the plantings are equally kept simple with mainly wisteria, privet, various grasses and slender Himalayan birch trees.

The sloping terrain has been made use of to provide sitting steps, and residents can take part in activities on multi-functional squares and spaces, which include a covered stage and special areas for dancing, skating and ball games. Four footbridges provide new icons for the area. Together with new, more secure lighting, they also help to provide landmarks in the night, each illuminated in its own colour.

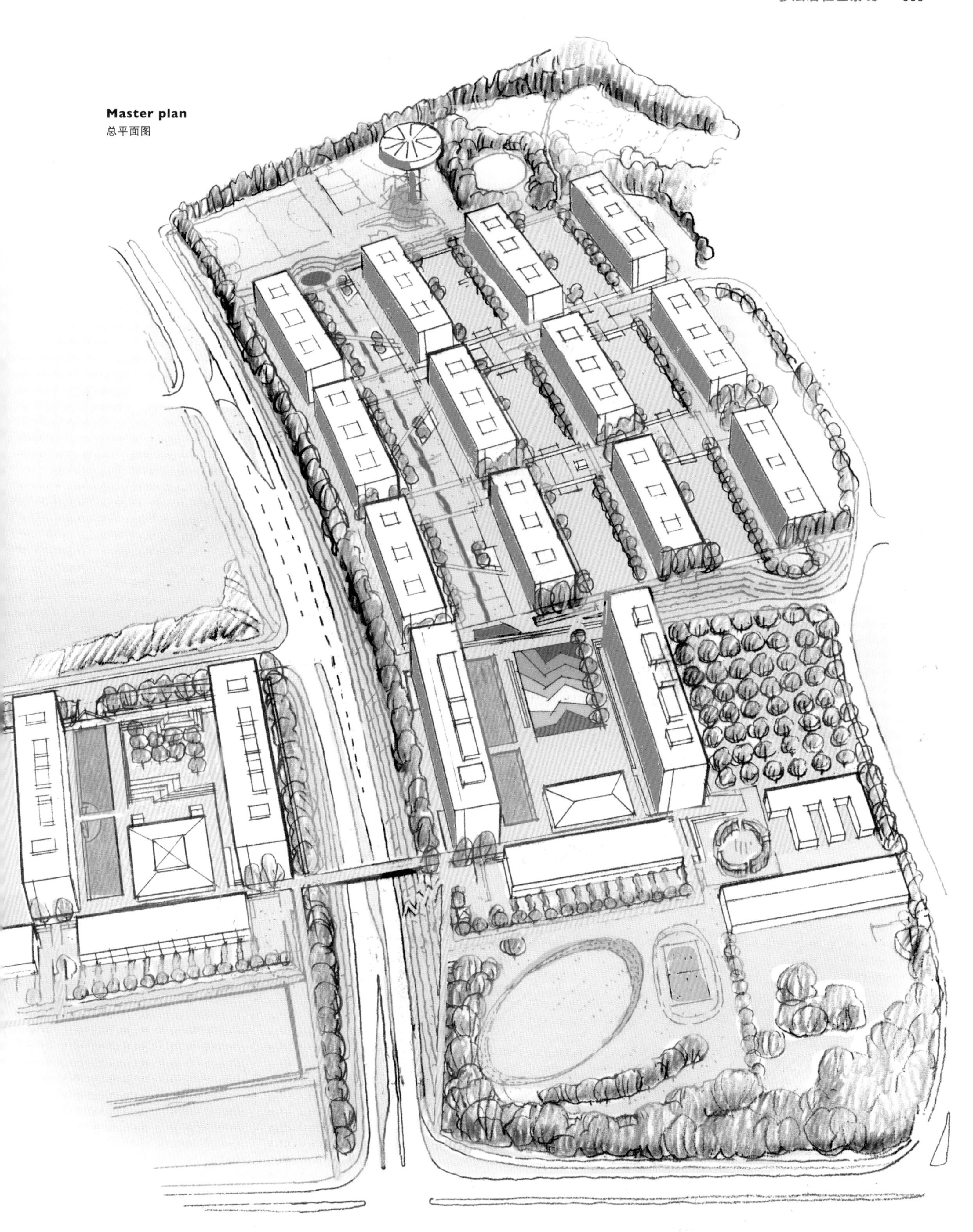

Master plan

总平面图

1. New meeting places for the residents
2-4. The open space is simple but comfortable

1. 全新的集会空间
2～4. 宽敞的户外空间虽然简单，但是看着很舒适

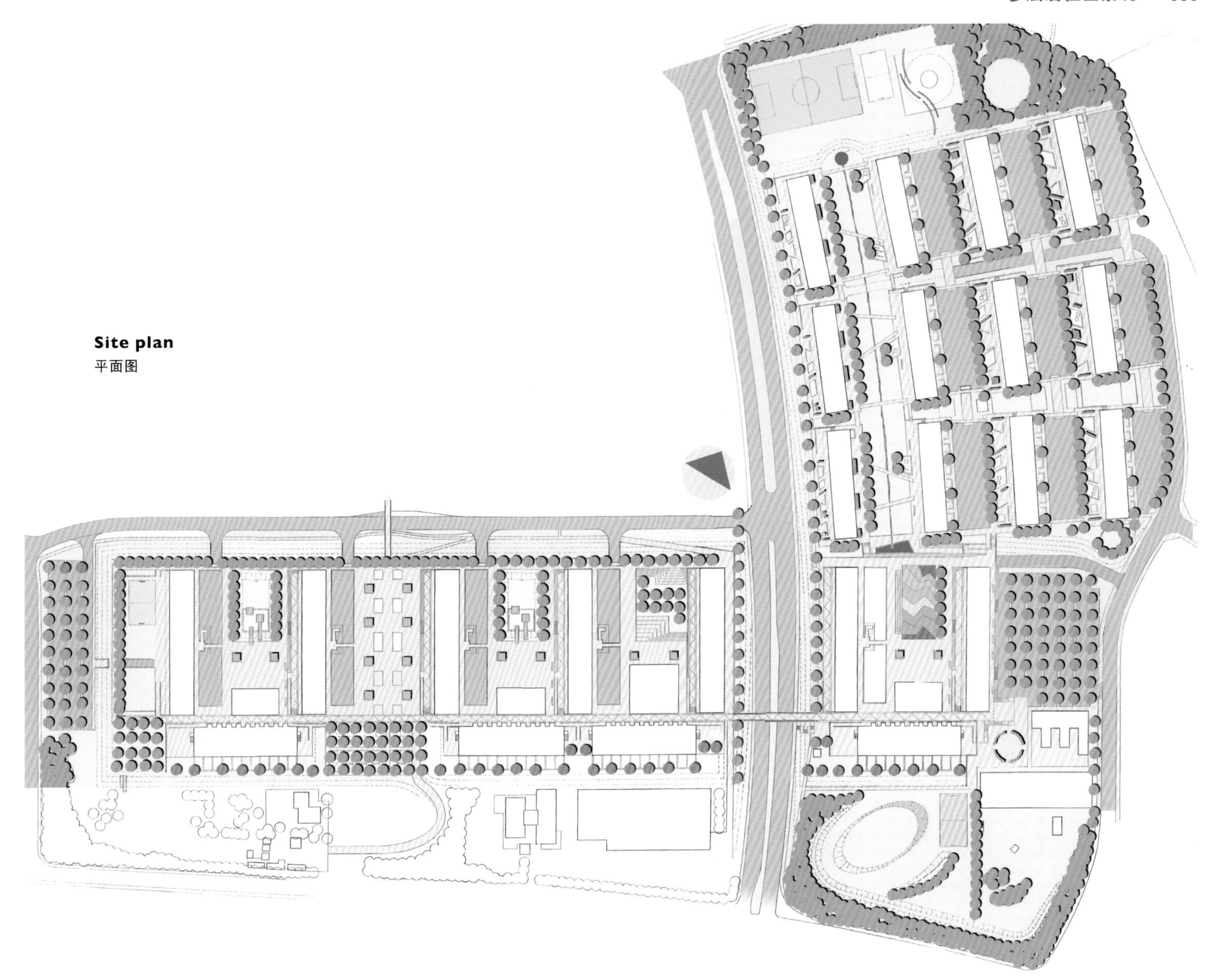

Site plan
平面图

比斯普海文住宅区建于1970年，住宅区的户外空间以大面积的裸露区域、高大的混凝土围墙和茂密的植物为特色，形成了单调且不安全的户外环境。

尽管拥有迷人的公寓和靠近市中心的地理位置，该地区给人以类似贫民区的形象，有许多破坏公物的行为，因此很难吸引新租户。整体翻修项目的目标是为该地区带来全新的形象和品质，特别是为居民营造若干个全新的集会空间。

设计师用草坪和花坛替代了原来茂密的植被，让户外区域看起来更开敞。环绕开发场地的独特斜坡上的灌木被清理干净，取而代之的是开放的草皮和成排的桦树。高大的混凝土围墙被切除顶端，贴近地面，形成了低矮的围栏。

5

6

7

5. The new paving uses a restrained palette of black and white patterns
6-7. Details of the stairs
8-9. The plantings are equally kept simple with mainly wisteria, privet, various grasses and slender Himalayan birch trees

5. 新的铺装采用低调的黑白图案
6、7. 台阶的细节图
8、9. 植被的栽种也尽量保持简单，以紫藤、水蜡树、各种草和细长的喜马拉雅桦树为主

新的铺装采用低调的黑白图案，与住宅楼翻新的立面相互搭配。植被的栽种也尽量保持简单，以紫藤、水蜡树、各种草和细长的喜马拉雅桦树为主。

设计充分利用了坡式地形来营造休息台阶，居民可以在多功能广场上进行活动。广场包括带顶舞台以及舞蹈、轮滑和球类运动专区。四座人行天桥成为了该地区的新标志，它们与全新的安全照明结合起来，在夜晚以四种不同的色彩指引着人们的方向。

LODENAREAL
伦德纳瑞尔住宅

Location: Innsbruck, Austria
Design: Monsberger Gartenarchitektur
Photography: Monsberger Gartenarchitektur
Area: 24,500sqm

项目地点：奥地利，因斯布鲁克
设计师：格特劳德・蒙斯伯格景观建筑事务所
摄影：格特劳德・蒙斯伯格景观建筑事务所
面积：24,500平方米

Combination of Urban and Landscape

The Stage – Window to the Landscape: The basic idea is to integrate the surrounding landscape as an identity feature. The increased grass platform takes on the role as mediator between the city (residential areas) and landscape (Inn) and leads the eye across to the mountains. Loosely strewn mountain ash is used as a landscape quote emphasises the character of this place.

Public Open Spaces

The Tape – Leisure and Recreation: The linear strip is zoned in the bank free from the natural environment in certain areas and certain urban and functional rooms. The immediate riparian zone is maintained as a gravel bank and open space and is flooded according to the water level temporarily. Stone body in the form of groynes provides direct access to the water at higher water levels. The dam is designed as an Inn promenade for pedestrians and cyclists. Chopped rear body structure in the south of the dam subsequent open space is formed as a function of volume (e.g. for beach volleyball, street ball, skating).

A single row of avenue planting (bird cherry) and the uniform planting of the embankment with sloes emphasises the linearity of the band and is interrupted only in the stage.

Free Space Residential Development

The Bar Code – Yellow, Red, Blue: Each courtyard is determined by a colour that gives it a distinctive character. Plants and materials in the basic colours yellow, red and blue colour determine the respective game. The change of seasons influenced the intensity of the colour in all its nuances.

The formal shape results from a free interplay of horizontal lines of the facade, like a bar code can be projected onto the surface area of the farms. Different areas give the code and allow a wide variety of uses.

1. Bird eye view
2. The increased grass platform takes on the role as mediator
3. Loosely strewn mountain ash is used as a landscape quote emphasizes the character of this place
4. The plants on the wooden deck

1. 鸟瞰
2. 新增的草坪扮演了调节者的角色
3. 铺撒的火山灰突出了该地区的景观特征
4. 木质甲板上种植的树木

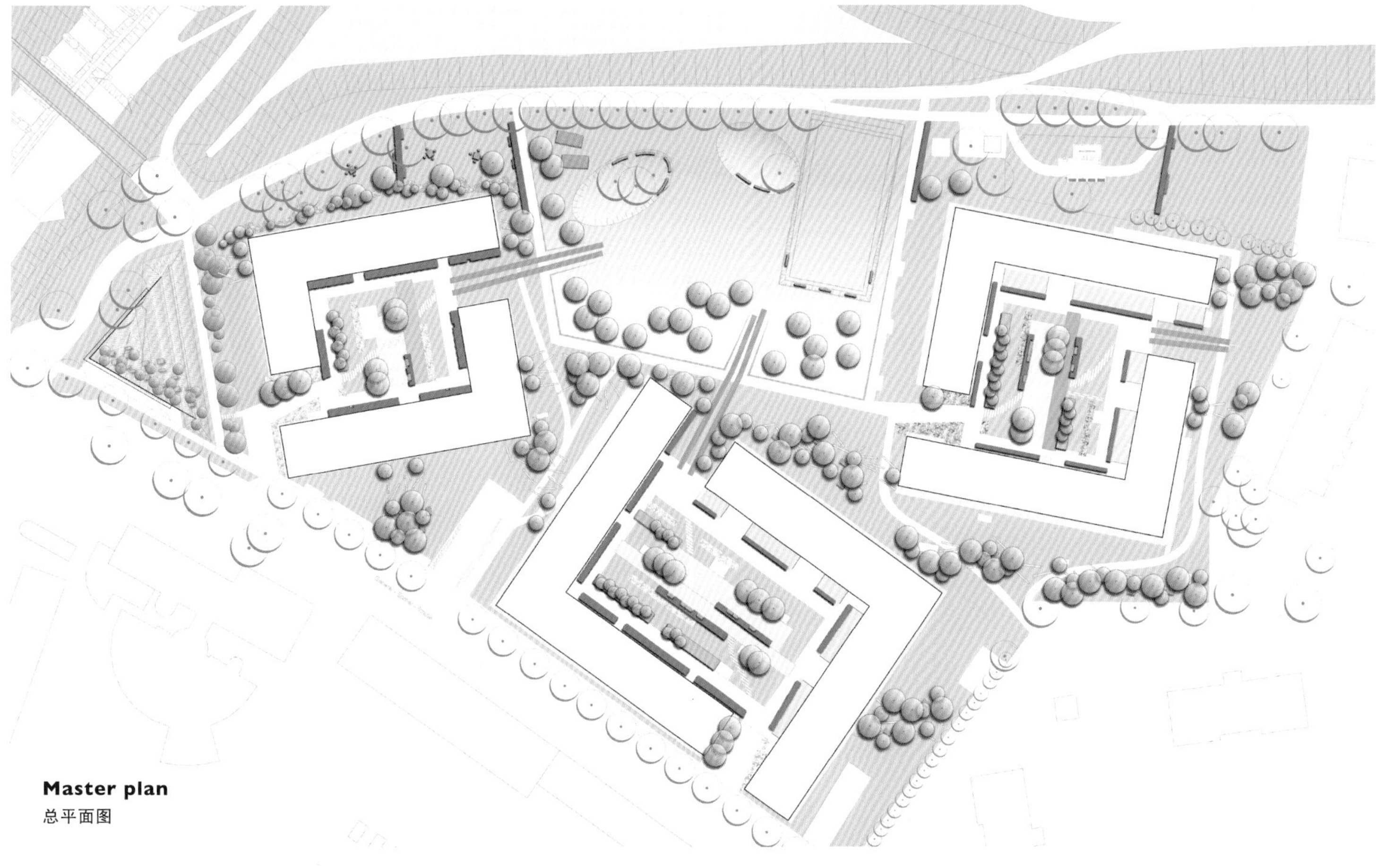

Master plan

总平面图

2

3

4

城市与景观的结合

舞台——窗口面向景观：项目的基本概念是将四周的景观融合为一个标志性特征。新增的草坪扮演了城市（住宅区）与景观（因河）之间传递者的角色，将人们的视线引向远处的高山。铺撒的火山灰突出了该地区的景观特征。

公共开放空间

景观带——休闲与娱乐：沿河景观带被划分成若干个自然区和若干个城市功能区。河岸区被保留成碎石河岸景观和开放空间，有时会因水位上升而被淹没。当水位较高时，石砌防波堤直通河水。水坝被设计成河畔长廊，供行人和骑行者使用。水坝南侧的后部被削平，形成了连续的开放功能空间，包括沙滩排球场、街头篮球场、轮滑场等。

单排行道栽植（稠李）和统一的堤坝栽植（黑刺李）突出了景观带的直线感，仅在开放空间处会被打断。

自由空间住宅开发

条形码——红、黄、蓝：每个庭院都有其独特的色彩。红、黄、蓝三原色的植物与材料决定了不同的游戏。随着季节的变换，庭院色彩的强度也会产生细微的变化。

景观的正式造型来自于建筑立面的水平线条，就像是投射在牧场地面上的条形码一样。不同的区域拥有不同的条码，能实现不同的功能。

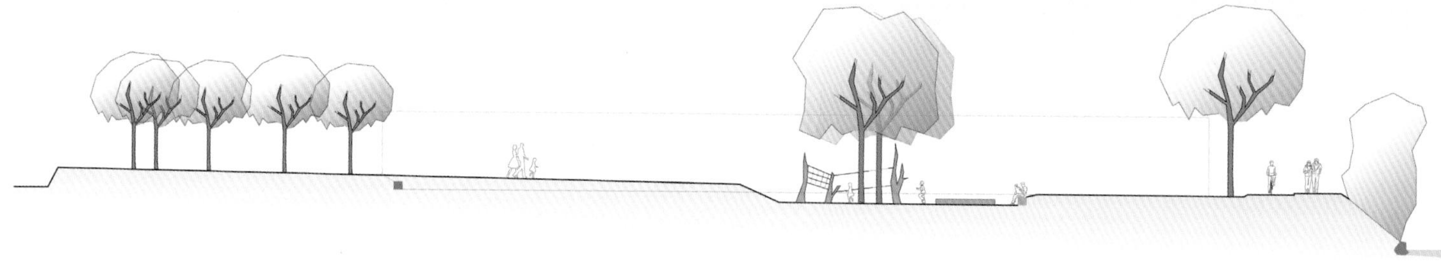

Concept drawing
设计图纸

5. Green lawn space for relaxing
6-7. The dam is designed as an Inn promenade for pedestrians and cyclists
8. The beach playground

5. 大片的草地是休憩的好去处
6、7. 水坝被设计成河畔长廊，供行人和骑行者使用
8. 沙滩游乐场

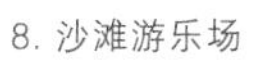

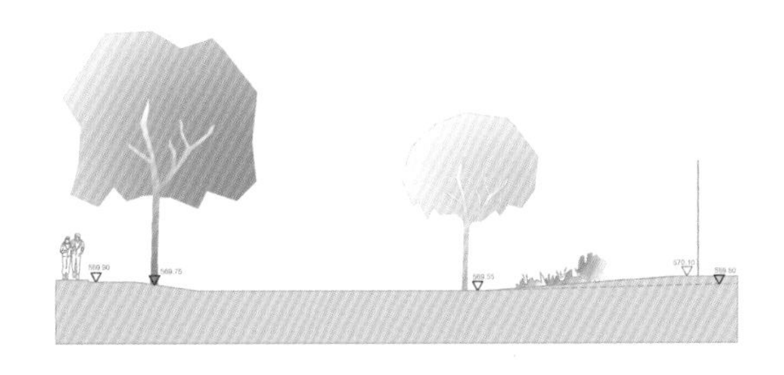

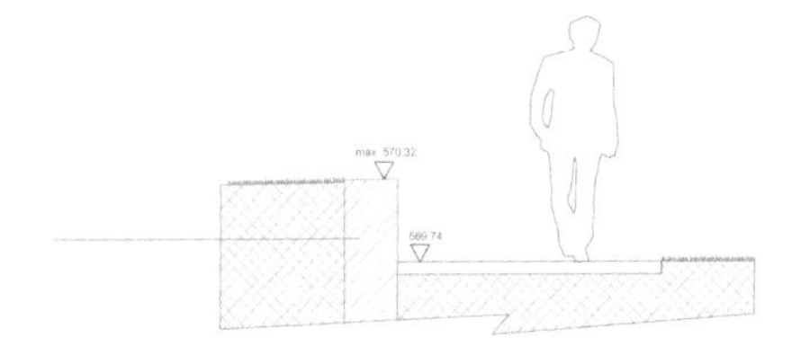

Sections
剖面图

CSU FULLERTON HOUSING PHASE III

加州大学富尔顿分校三期住宅

Location: Fullerton, CA, United States
Completion: 2011
Design: CMG: Landscape Architecture
Architect: Steinberg Architects
Photography: CMG: Landscape Architecture
Area: 36,423sqm

项目地点：美国，富尔顿
完成时间：2011年
设计师：CMG：景观建筑事务所
建筑师：斯坦伯格建筑事务所
摄影：CMG：景观建筑事务所
面积：36,423平方米

CMG collaborated with PCL Construction and Steinberg Architects for the design of this wining student housing competition entry. The extensive design/build proposal encompasses a long list of services in addition to student housing. Included in the proposal were student coordinator and faculty-in-residence apartments, administrative offices, conference and multi-purpose rooms, laundry and mail facilities, recreational lounges, a convenience store, a maintenance facility, a central plant, and a dining facility, along with 4 acres of new campus open space amenities on an 8.63 acre site. The landscape accommodates student activity of all scales, binds the new construction together as a district, and affords easy connections to the existing campus fabric.

The new heart of the campus housing district provides a large, flexible event space as well as a multitude of smaller gathering spaces for outdoor classrooms, study, and relaxation in the shade of broad canopy trees. All plant material is climate-appropriate, and stormwater quantity and quality control measures are fully integrated into the site design. The design team intends for the project to be LEED certified.

CMG景观设计公司与PCL建筑公司和斯坦伯格建筑事务所共同完成了学生公寓的设计。除了公寓楼之外，项目还融入了大量附加的服务设施。整个项目包含：教职工及学生公寓、行政办公室、会议室及多功能厅、洗衣房和收发室、娱乐休息室、便利店、维护设施、中央电站、餐厅以及16,187平方米的全新校园开放空间。景观设计满足了学生各个方面的活动需求，将新建设施整合成一个整体，在已有的校园网络中添加了便捷的道路连接方式。

校园住宅区的新中心提供了宽敞而灵活的活动空间以及各种各样的小型户外教学、研究、休闲空间，并且以茂密的树木为这些空间提供了阴凉。设计所选用的所有植物都能适应当地的气候。整个场地设计添加了暴雨控制措施。设计团队决定为项目申请绿色建筑LEED认证。

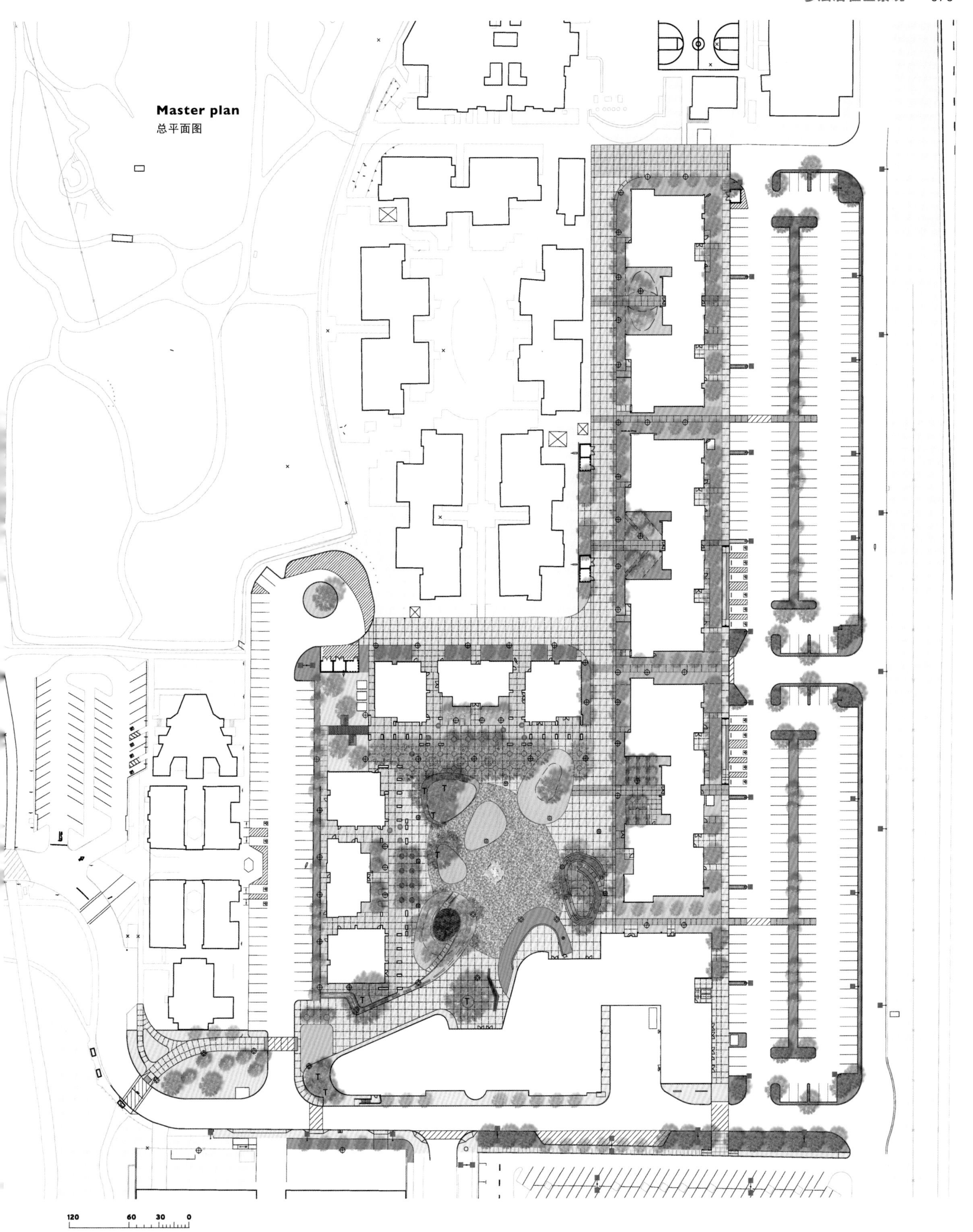

Master plan

总平面图

1-2. The colourful pavement
3. Succulent plants
4. The fountain

1、2. 五颜六色的特色铺装
3. 多肉植物
4. 喷泉

3

4

5. The stone seatings
6-7. The wooden seatings
8. The plants chosen are climate appropriate

5. 石质座椅
6、7. 木质座椅
8. 所选植物都是能适应当地气候的

8

ROCKBROOK RESIDENTIAL COMPLEX

石溪住宅区

Location: Dublin, Ireland
Design: Bernard Seymour Landscape Architects
Photography: Derek Naughton
Area: 1,200sqm
项目地点：爱尔兰，都柏林
设计师：伯纳德·西摩景观建筑事务所
摄影：德里克·诺顿
面积：1,200平方米

This is a residential block, part of a much larger scheme, involving the master-planning and development of a district adjacent to Beacon South Quarter in Sandyford. The scheme is reasonably tall for the area, seven floors at its maximum, with two floors of underground car-parking. The challenge was to make a space of quality and utility for the residents where the sensation was of a courtyard in the most conventional sense, as opposed to some token greenery placed atop the car-park.

The concept was based on a woodland clearing, where one emerges from the shadows into a dappled area, where one can take advantage of the sunlight and warmth by resting a while on a fallen log. This fanciful notion had to be fused with the need to provide a useful amenity for the residents as the only communal space that they would all share.

A social space was desirable, with the conventional elements that might be expected, such as grass, seating, and a gathering area where residents might occasionally meet and put out some chairs and tables. Some greenery provided seasonality, scale and softness. The tricky part was not to make it look like planters placed above an underground car park, which is the reality of the project.

石溪住宅区是爱尔兰首都都柏林桑迪福特区附近一个街区总体规划开发的一部分。小区内最高的住宅楼为七层，在该地区属平均水平，下设两层地下停车场。设计所面临的挑战是为居民打造一个富有庭院感的高品质户外空间，而不是像某些建在停车场上方的象征性绿化一样。

设计概念以“林间空地”为基础，人们可以在这里感受斑驳的光点，在倒下的树干上享受阳光与温暖。这一概念必须与使用的居民娱乐设施结合起来，因为这是他们在小区里唯一可共享的公共场所。

一个令人满意的社交空间应当具有草坪、休息区、供居民偶尔会面的集会区以及桌椅灯传统元素。一些绿色植物能为空间带来四季的变化，让整个空间变得融合。设计的重点不是让景观看起来像是摆放在地下停车场之上的花池，而是一个高品质的庭院。

Master plan
总平面图

1

1. Roadside, with designed ventilation grills
2. Overview
3. Bench with shadow gap

1. 路边的特色通风格栅
2. 总体效果
3. 休闲长椅

2

3

4
5
6
7

4. Path	4. 人行路
5. Dynamic lines	5. 动感线条
6. Material detail	6. 材料细节
7. Cyclist	7. 骑自行车者
8. Variety of planting	8. 多种多样的植栽
9. Red Hot Pokers	9. 叶兰
10. Flowerbeds	10. 花床
11. Wheelchair ramp engulfed by vegetation	11. 轮椅坡道种满了植物

SUN CITY YOKOHAMA

横滨太阳城

Location: Yokohama, Japan
Design: SWA Group
Photography: Tom Fox
Area: 75,000sqm
项目地点：日本，横滨
设计师：SWA集团
摄影：汤姆·福克斯
面积：75,000平方米

Sun City Park Yokohama is a new CCRC (continuum of care retirement community) proposed for a site that had formerly been an explosives manufacturing facility in Hodogaya, a suburb of Yokohama, Japan. The project, to be operated by Health Care Japan Co., Ltd. (HCJ), a leader in Japan's fast growing senior housing and continuum care retirement communities, adds to an expanding portfolio of properties focused on housing Japan's growing senior population. This 7.5 hectare, former explosives manufacturing facility overlooking Yokohama provided a unique opportunity to allow the design team to formulate a master plan sensitive to the natural character and charm of this very unique site. The hilltop setting, woodland edges, and sensitive building layout, in conjunction with the introduced landscape present Sun City Yokohama as a landmark senior community in Japan.

With Perkins Eastman Architects PC, Pittsburgh, PA, SWA has completed the master planning phase of the project and is currently proceeding with schematic design. The plan consists of two single building "villages" connected by a pavilion-like community building. Each village has 240 Independent Living (IL) units, each with its own community living and dining programs. The west village also contains the 120-bed skilled nursing facility with its own arrival court on the north side of the building. The community building spans a natural draw in the landform that with the east village frames a large meadow that rolls toward a created stream that runs along the toe of a steep tree covered slope that forms the west edge of the space.

The residential wings of the villages extend into the landscape offering views into gardens, woodland preserve, and stunning distant views to Yokohama City to the east. The building composition thus frames a large stroll garden in the valley viewed from the main public rooms of both villages and residential units above, features a large open lawn and woodland understory garden separated by a gently cascading stream. For seniors, the compelling views of "garden" and nature from the buildings are an important part of everyday life, and perhaps as important as being in the landscape itself. The preserved and created landscapes of Sun City Park Yokohama meet that goal and more with a variety of outdoor spaces to accommodate various

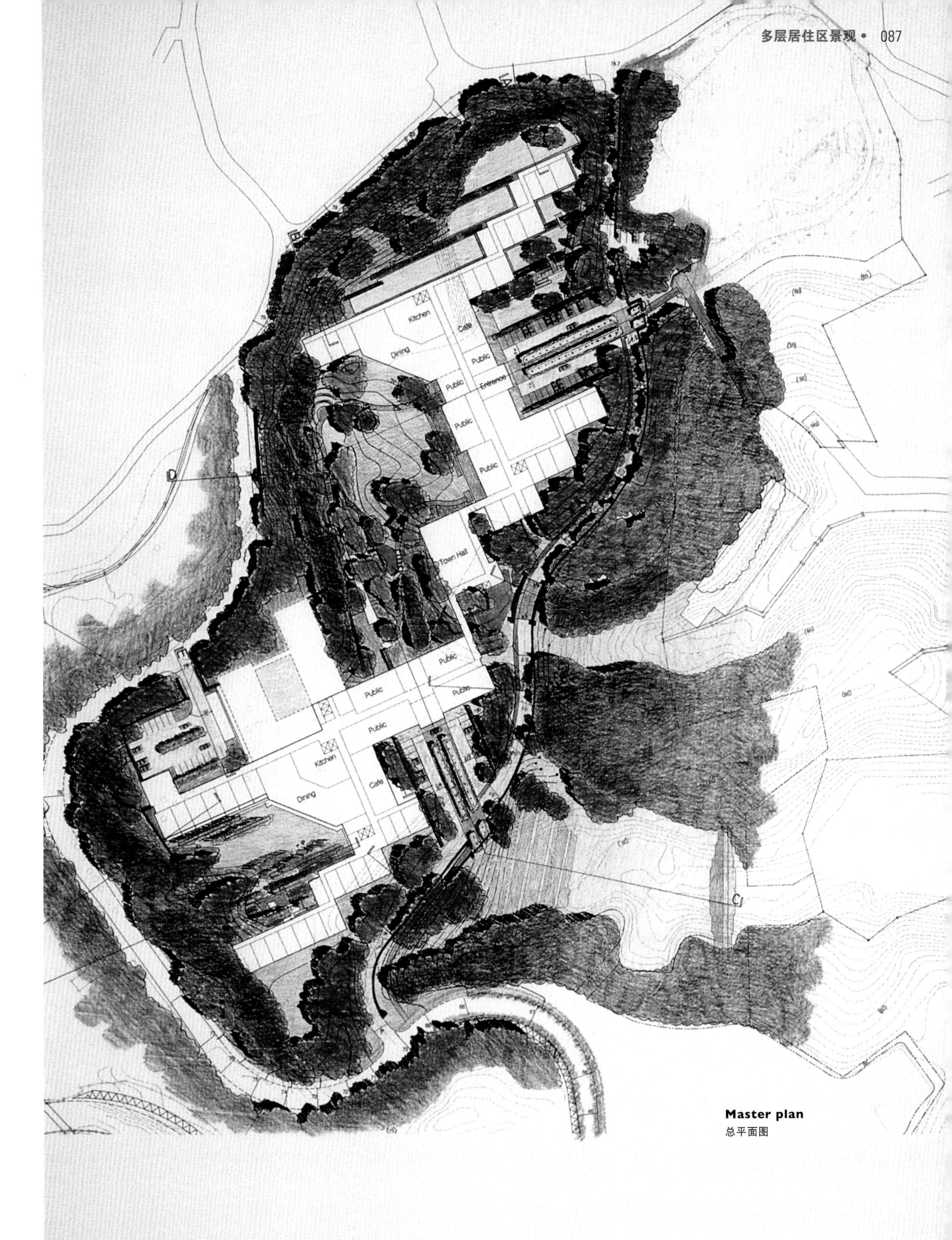

Master plan
总平面图

1. Colourful flowers are planted
2. The community is like a garden
3. The path is wandering in the green plants

1. 五彩缤纷的植物
2. 花园式的社区景观
3. 幽静的小路在绿色植物中

needs of the elderly including more active residents, as well as, larger group gatherings.

SWA's involvement in this project began with a feasibility study of the site and a review of the existing development scheme proposed by a competing senior housing developer. The parcel size and context provides the opportunity to build a facility twice the size of any project in a truly park-like setting. The feasibility study confirmed that the previously proposed and approved unit count could be achieved without approaching the project as a terraced hillside development.

The greatest challenges for SWA are to sensitively grade the edges, to save existing mature trees and to design the project to maximise the use of existing site amenities. The entry road set below the pad elevation runs along the natural and partially wooded edge that will be reinforced to complete the "...through the woods to Grandmother's house we go" image.

横滨太阳城是一个新式持续护理退休社区，建在日本横滨城郊保谷区一座炸药制造厂的原址上。项目由日本发展最迅速的老年住房和持续护理退休社区运营公司——日本医疗保健公司运营，该公司主要关注日本快速增长的老年人口。占地7.5公顷的前炸药制造厂场地为设计团队提供了打造适应该地区自然特征和独特魅力的整体规划的机会。山顶环境、林地边界以及敏感的建筑布局与景观元素结合起来，让横滨太阳城变成了日本老年社区建设的典范。

SWA集团与建筑师珀金斯·伊士曼建筑事务所共同完成了项目的总体规划。项目由两个独立的疗养村组成，二者通过一座公共楼连接起

2

3

4. The unique pavement
5. Concise and neat landscape
6. Seating place for relaxing
7. Green plants are good for relaxing

4. 独具特色的铺装
5. 简洁干净的景观设计
6. 供人休憩的室外座椅
7. 绿色植物适宜身心放松

7

来。每个疗养村拥有240套独立住宅单元，分别配置社区生活项目和就餐设施。西侧疗养村还包含120个特殊护理床位，在建筑北侧有独立的到达庭院。公共楼随着地势自然延伸，与东侧疗养村共同围起了一大片草地。草地向人工小溪延伸，小溪沿着种满树木的陡坡顺流直下，形成了空间的西侧边缘。

疗养村的住宅翼楼延伸到景观之中，享有花园、林地和远方横滨城的景色。建筑组合在山谷中围出一个巨大的漫步花园，花园以巨大的开放草坪和林地下层植被花园为特色，二者被一条顺流直下的小溪隔开。两个疗养村的主要公共空间和上方的住宅单元都能享受这些美景。对老年人来说，拥有花园和建筑周围自然环境的视野是日常生活的重要组成部分，甚至可能与这些景观本身同样重要。横滨太阳城的自然景观和人工景观实现了这一目标，并且以多样化的户外空间满足了老年人的各种活动乃至集会需求。

SWA集团对项目的参与始于对场地的可行性研究以及对竞争开发商所提出的开发规划的评估。项目地块和环境赋予了项目超凡的规模，让设计师可以打造一个真正的公园环境。项目的可行性研究肯定了之前的提案，认为项目无需在山坡上进行阶梯式开发就能实现预期的住宅单元数量。

SWA集团所面临的最大挑战就是如何实现空间的层次划分、如何保留已有的成熟树木以及如何最大化地利用已有的场地设施。山脚下的入口通道穿过大片的树林，将被打造成“穿过树林去往外婆家”的感觉。

8. A created stream runs along the toe of a steep tree covered slope
9. A large meadow that rolls toward a created stream
10. Sensitive building layout is in conjunction with the introduced landscape
11. The ornamental landscape lighting

8. 小溪沿着种满树木的陡坡顺流直下
9. 草地向人工小溪延伸
10. 建筑布局与景观元素结合起来
11. 具有装饰性的景观照明

9

10

11

GREENCIA KAWASAKI-KYOMACHI

GREENCIA川崎京町

Location: Kawasaki, Japan
Completion: 2012
Design: Hiroshi Ishii (Yoshiki Toda Landscape & Architect Co.,Ltd.)
Photography: Yoshiki Toda Landscape & Architect Co.,Ltd.
Client: NIPPON STEEL KOWA REAL ESTATE CO.,LTD., Nice Corporation, HASEKO Corporation
Area: 14,587.48sqm

项目地点：川崎市，日本
完成时间：2012年
设计师：株式会社户田芳树风景计画；石井博史
摄影：株式会社户田芳树风景计画
委托人：新日铁兴和不动产株式会社（KOWA REAL ESTATE）、Nice株式会社、株式会社长谷工 Corporation
面积：14,587.48平方米

1. Entrance Garden – overlook of landscape axis at the entrance from main feature tree
2. Street Garden – landscape wall and small terrace in linear direction which provide space variation

1.入口庭园，从主景树处眺望主入口处的通景线
2.街道庭园，直线方向上赋予空间变化的景墙与小型平台

Landscape in a Grid

Located in Kawasaki-Kyomachi, Japan, GREENCIA Kawasaki-Kyomachi is an apartment development with 360 units. Now the block looks clean and neat, and the residential buildings combine with the park with natural landscape harmoniously, creating a calm and peaceful ambience. In a human-scale neighbourhood with over 100 years history, a lush apartment landscape of nearly 1.5 hectares which makes the best of the favourable conditions unfolds itself to the public.

Housing in Garden

It is noticeable that the project connects the enclosing spaces around the buildings and all the open spaces are open to public. If we say that residential buildings are everyday living places, outdoor spaces play an equal important role. The proposal aims to treat the surrounding outdoor space as a lush garden, through which man-to-man and man-to-nature direct contact happen. The ultimate goal is to create "Housing in Garden".

Six Landscape Gardens

The site is somewhat special because the distance from the front gate to the main entrance is about 180m. The architectural composition incorporates the irregular site and the main challenge of the preliminary phase is how to lead people to the main entrance area through landscape design. Generally speaking, the longer home-entry distance makes people feel alienated. However, with proper design, this disadvantage can create a charming path. In the programme, the landscape designers first apply some feature trees and rows of landscape trees at the end of the road leading to the front gate, which reduce people's resistance to distance through specific goals. Then, constant scene changes along the path create a pleasant pedestrian space. Meanwhile, seasonal plantings along the path enhance the landscape effect. In addition, the continuous outdoor space is divided into six areas. With different themes, the six gardens improve the circulation of the whole site and create continuous landscape in collocation with the terrain.

1

2

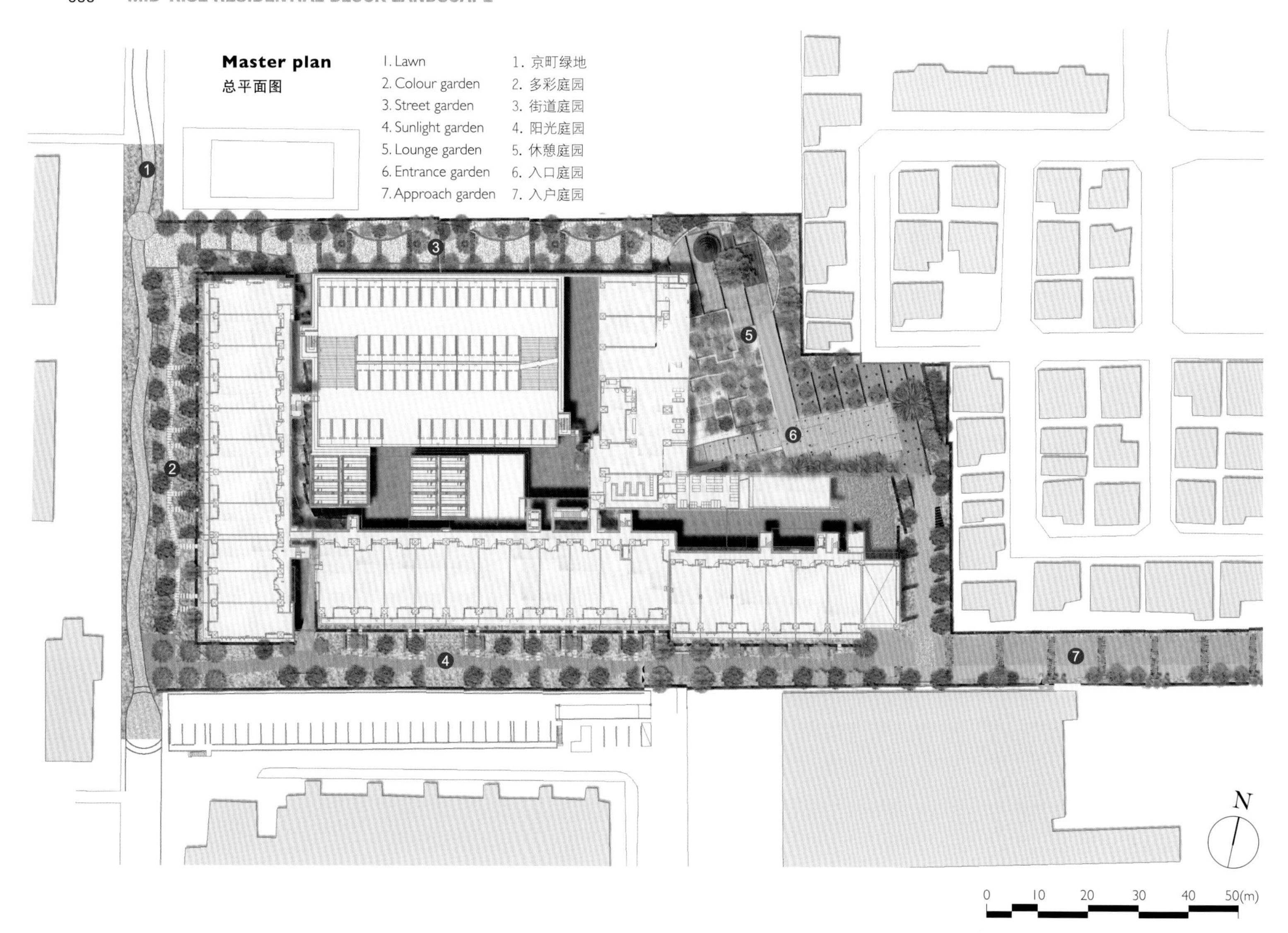

From Ornamental Garden to Living Garden

The gardens are not ornamental. Here, everyone can contact with nature and explore their own stories. The result is a living garden. The designers hope, in this delightful garden housing complex, human being, vegetation and other creatures will grow vigorously and here will become a valuable local landscape resource for the public to share.

3. Approach Garden – entrance columns and Cercidiphyllum border trees

3.入户庭园，迎接访客的列柱入口空间与连香行道树

棋盘格中的景观

GREENCIA川崎京町是位于日本川崎市川崎区京町的360户出售公寓。现在这里的街区齐整，住宅楼与多自然风景的公园错落有致地结合在一起，营造出祥和恬静的氛围。在延续了百余年的人性化尺度街区内，一个面积约1.5公顷，充分利用了用地优越条件、绿意丰盈的公寓景观展现在人们面前。

庭园里的住宅

项目设计中值得一提的，是将围合在建筑周边的外部空间衔接起来，其中任何一块空地都面向市民开放。若说住宅楼是每天利用的生活场所，那么户外空间也发挥了同样重要的作用。本次规划的目标，是将围绕建筑的户外空间视为一个丰盈润泽的大庭园，通过这里人与人、人与自然的零距离接触，最终营建出“庭园里的住宅”。

六个庭园物语

本项目用地形状较为特殊，从正门到住宅区主入口的距离约180米。建筑布局完全结合了不规整的用地，而景观上如何将人们引导至住宅主入口区域成为规划初期所面临的主要课题。通常入户距离越长越让人“敬而远之”，但如果采用合理的设计方式，也能将冗长的道路变得魅力十足。在本次规划中，首先在正门前方的尽头设置了具有视线停留功能的主景树和景观树列等，通过赋予行人明确的目标来缓解人们对长距离产生的抵触情绪。其次，在到达尽头的过程中利用间歇性的场景切换，营建出令人愉悦的步行空间。同时，在路边种植了四季盛开不断的花草等元素，增强狭长空间的景观效果。另外，将户外连续性空间分成六个主要区域，通过不同主题的六个庭园来提升用地整体的环游性，配合用地特殊形状营建出连续性景观。

从观赏的庭园到生活的庭园

这里不单纯是供人们观赏，每个人都能在此亲近自然，寻求属于自己的故事，最终使本项目成为“生活的庭园”。衷心期待着在舒适的庭园式住宅区中，人、植物及其他生物都能茁壮地成长，让这里成为当地宝贵的景观资源与大家分享。

4. Street Garden – waterside walkway which corresponds to curving landscape wall
5. Colour Garden – maple-based plantings and lawn create a soft image for the garden
6. Entrance Garden – shadows of border trees on the paving reflect the passage of time
7. Sunlight Garden – abundant sunlight and tree shade form a sunny grove

4. 街道庭园，与曲线造型修景墙相映衬的汀步散步道
5. 多彩庭园，槭树类为主的植栽及草坪地形营造出柔和印象的庭园
6. 入口庭园，落在铺装面上的景观树阵的树影反映出时间的推移
7. 阳光庭园，充裕的阳光与成片的树荫构成风和日丽的树林

8. Lounge Garden – terraced garden which senses time from water pond, gravels and groundcover vegetation
9. Stone-faced feature wall of Approach Garden – main feature trees standing deep behind the rhythmic landscape walls
10. Entrance Garden – Cercidiphyllums and main feature trees at the main entrance
11. Entrance Garden – needle-leaved tress at the entrance

8. 休憩庭园，从水盘、砂砾、地被类植物中感知时间推移的阶梯式庭园
9. 入户庭园的石贴面景墙，在富有节奏感景墙深处耸立的主景树
10. 入口庭园，承接从主入口投来视线的连香树树阵与主景树
11. 入口庭园，承接来自主入口视线的针叶主景树

AVALON OCEAN AVENUE

阿瓦隆海洋大道

Location: San Francisco, USA
Completion: 2012
Design: Jeffrey Miller (Miller Company Landscape Architects)
Photography: Miller Company Landscape Architects
Area: 30,263sqm
项目地点：美国，旧金山
完成时间：2012年
设计师：杰弗里·米勒（米勒景观设计公司）
摄影：米勒景观设计公司
面积：30,263平方米

Avalon Ocean Avenue is a mixed-use residential and commercial development in the Balboa Park neighbourhood of San Francisco. This transit-oriented development is served by improved light rail and bus service along Ocean Avenue, as well as by the nearby Balboa Park BART station. The Ocean campus of the City College of San Francisco and the Ingleside branch of the San Francisco public library are immediately adjacent. Street-level retail includes a Whole Foods Market.

A pedestrian-oriented streetscape has been created along Ocean Avenue. Brighton Street extends into the development as an active "Woonerf" curbless street, with a raised terrace area that includes concrete seat walls and other pedestrian amenities. The street provides vehicles access to underground parking structures.

Residents can enter the building from the parking structure or through the street entry, which is flanked with palms in raised planters. Podium courtyards within the buildings create communal and private outdoor space for residents, with raised concrete planter boxes and a variety of fixed and movable seating.

The west podium features a wavy concrete bench and an outdoor cooking area with barbeques, a sink, and counters. Residents can enjoy a sunny day in the communal spaces, or sit in private patios on the ground level.

The east podium also allows for outdoor cooking and dining. It includes a dedicated space for gatherings with tables and chairs, as well as two comfortable seating areas around a fireplace. Raised planters provide some privacy and enliven the central space.

1. Patio east podium
2. Podium courtyards within the buildings
3. Raised concrete planter boxes
4. A variety of fixed and movable seating
5. A dedicated space for gatherings with tables and chairs

1. 东部露天平台
2. 建筑内部的小庭院
3. 混凝土花池
4. 各种形式的座椅
5. 精致的集会空间，摆放着桌椅

Site landscape plan

景观总体平面图

1. Residential lobby
2. Residential lobby
3. Ocean Avenue
4. Commercial lobby

1. 住宅大厅
2. 住宅大厅
3. 海洋大道
4. 商业大厅

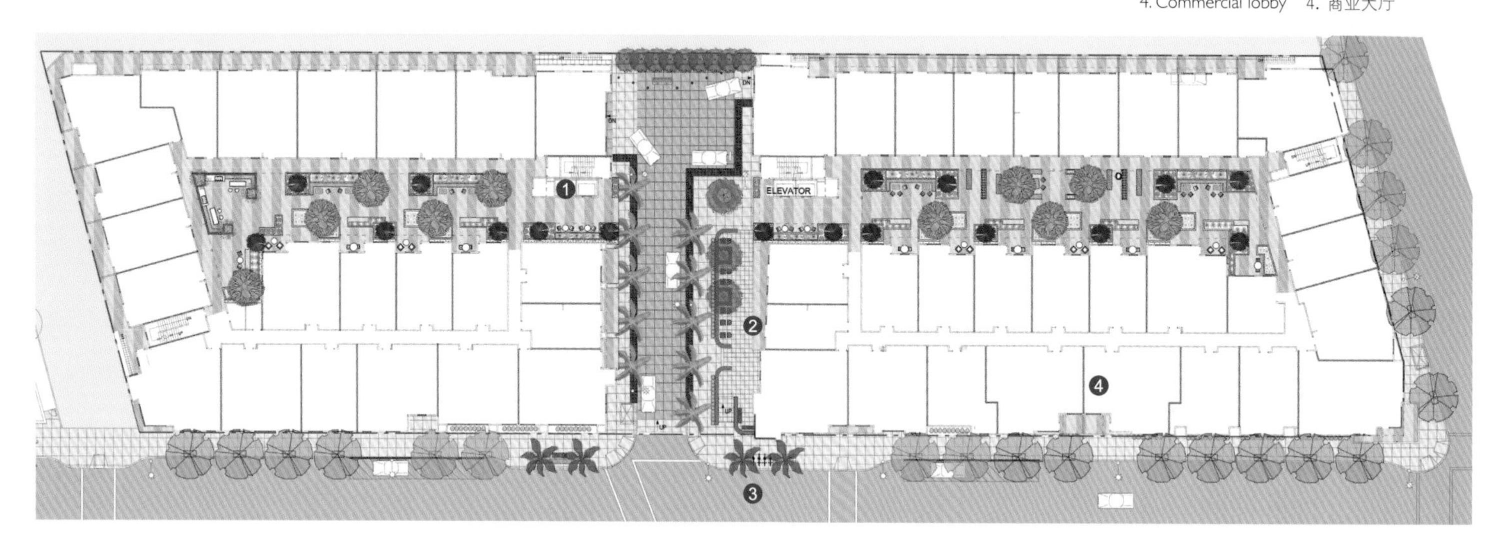

3

阿瓦隆海洋大道是一个商住混用项目，位于旧金山的巴尔博亚公园附近。这个以公共交通为导向的开发项目享有海洋大道沿线的改良轻轨和巴士服务，同时还靠近巴尔博亚捷运车站。旧金山城市学院的海洋小区以及旧金山公共图书馆的英格赛分馆都位于项目附近。街道层面的零售店则包括一家全食超市。

设计师沿着海洋大道打造了一系列以步行为导向的街景。布莱顿街一直延伸到项目内部，形成了一条活泼的庭院式无路缘道路。抬高的平台区内包含混凝土座椅墙和其他步行设施。街道还为地下停车场提供了机动车入口。

居民可以从停车场或街面入口进入建筑，街面入口两侧的花池中栽种了棕榈树。建筑内部的小庭院通过混凝土花池和各种形式的座椅为居民提供了公共和私密的户外空间。

西侧庭院以波浪形混凝土长椅和露天烹饪区为特色，烹饪区设有烧烤架、水槽和料理台面。居民可以在公共空间里尽情享受阳光，或是坐在一层的私人中庭中休息。

东侧庭院同样可以进行露天烹饪和就餐。它包含一个精致的集会空间，摆放着桌椅，还围绕着火炉设置了一个舒适的休息区。抬高的花池既保证了私密感，又活跃了中央空间。

4

5

6

6. The west podium features a wavy concrete bench and an outdoor cooking area with barbeques
7-8. Residents can enter the building from the parking structure or through the street entry

6. 西侧庭院以波浪形混凝土长椅和露天烹饪区为特色
7、8. 居民可以从停车场或街面入口进入建筑

VIA BOTANI

VIA波塔尼公寓

Location: Bangkok, Thailand
Completion: 2013
Design: Wannaporn Suwannatrai (Openbox Company Limited)
Photography: Wison Tungthunya
Area: 3,200sqm

地点：泰国，曼谷
完成时间：2013年
设计师：瓦那邦·苏万纳特来（OPNBX设计公司）
摄影：维森·东新雅
面积：3,200平方米

VIA Botani is a medium size condominium at Sukhumvit Soi 47, which is considered to be right at the heart of Bangkok. The site is in a tranquil neighbourhood of private housing development from early time. One thing that contributes the most to the unique site character is a fully-grown, magnificent rain tree that stands tall in the middle of the land. This rain tree is where the whole story begins.

At first glance, design team agrees that the rain tree must be kept as a living feature to ensure success of the development. Since then, design process and development revolves intensely around the rain tree; starting from feasibility study, building layout all the way through construction process. Keeping a large size tree alive and well at the center of construction work is extremely difficult, but well worth it. In the end, the tree stands proudly as heroic-scaled, approaching feature that everyone will remember.

Another part of the landscape work is the 2nd courtyard that consists of swimming pool and pool deck area. Following the architectural façade concept, "Illusion", landscape features arise from illusive lines, forms, patterns, and grow into an intensive and colourful courtyard; fully functional, yet very pleasing to the eyes.

OPNBX's philosophy is about harmonising Architecture and Landscape. Although the architecture scope is by another design firm, close coordination unite the concept of building and surrounding into one, pursuing the same concept more powerfully. It is the concept that grows into value, the value of conserving one single tree, VIA Botani.

1, The architectural façade concept arise from illusive lines, forms, patterns, and grow into an intensive and colourful courtyard

1. 建筑外墙的设计以令人产生错觉的线条、造型、图案为基础，形成了丰富多彩的庭院

1

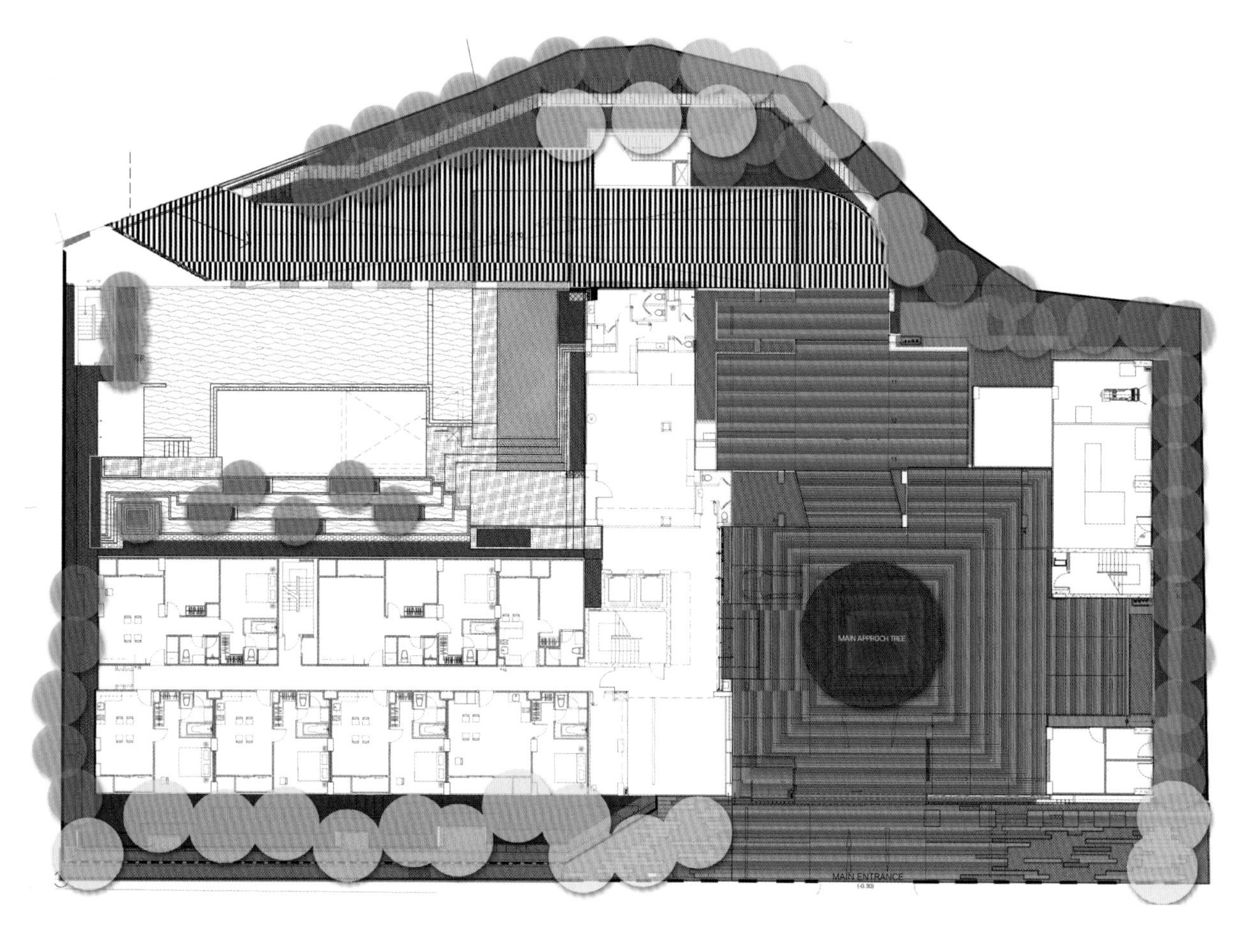

Landscape plan
景观平面图

VIA波塔尼公寓是一处位于维特索伊大道47号的中型公寓小区，位于曼谷的市中心。场地早前是一处宁静的私人住宅。项目的特别之处在于在场地中央有一棵高大的雨树。整个项目都围绕这棵雨树展开。

一开始，设计团队就认为这棵雨树必须被保留下来。整个设计流程和开发都紧密围绕着这棵雨树展开，从可行性研究、建筑布局，一直到施工过程。在施工场地中央妥善保留一棵鲜活而高大的树木并非易事，但是这一切都是值得的。最终，这棵树木成为了令人过目不忘的景观特色。

2. One thing that contributes the most to the unique site character is a fully-grown, magnificent rain tree
3. The featured green stairs
4. The tree pool
5. Paving detail

2. 项目的特别之处在于在场地中央有一棵高大的雨树
3. 绿色的特色铺装
4. 树池
5. 铺装细节

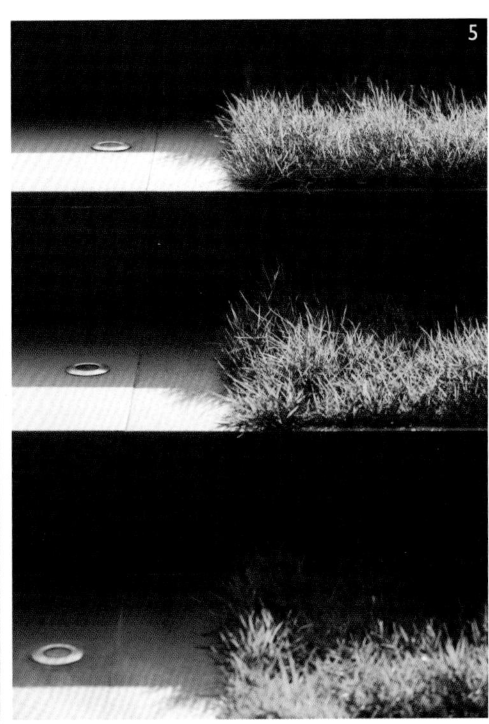

6

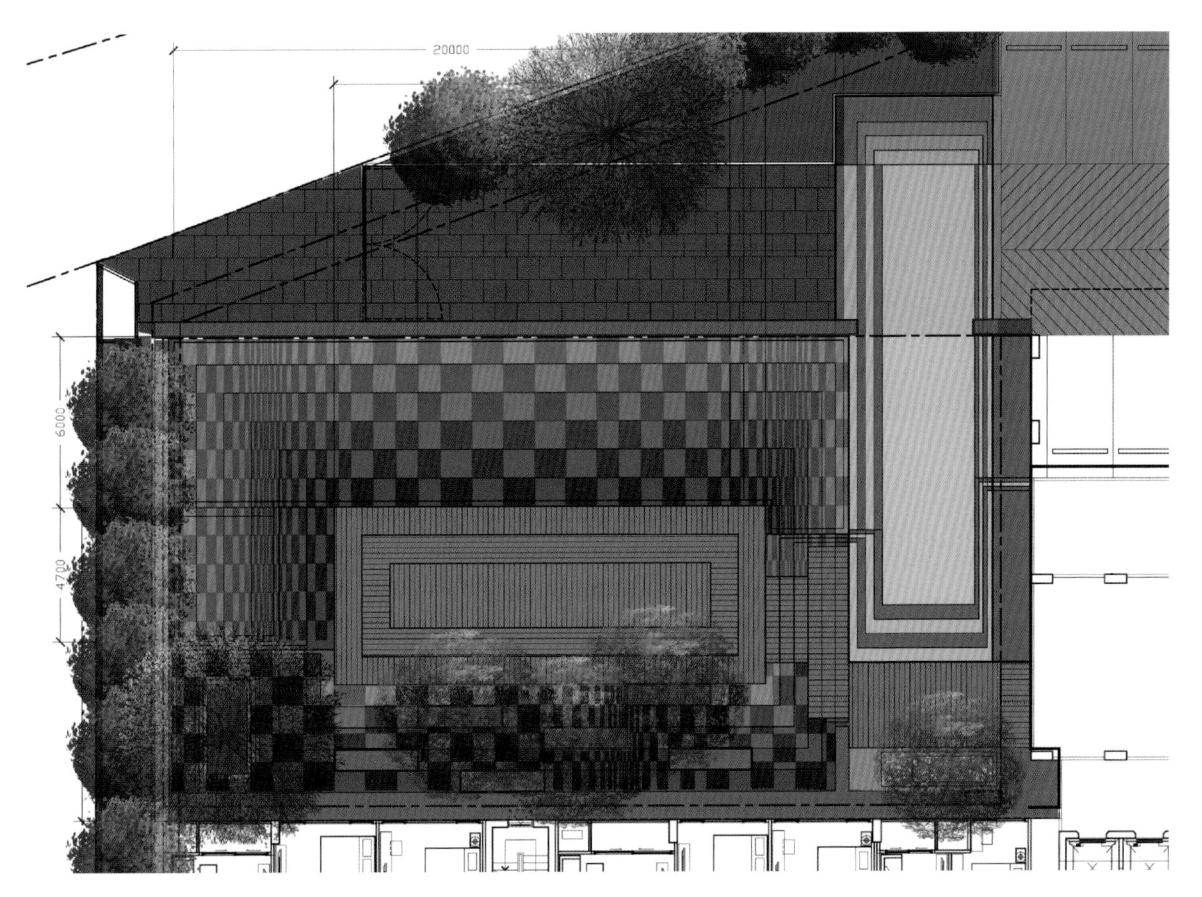

Swimming pool plan
游泳池平面图

景观设计的另一部分是由游泳池和泳池平台区所组成的第二庭院。景观设计延续了建筑外墙的设计概念“错觉”，以令人产生错觉的线条、造型、图案为基础，形成了丰富多彩的庭院，既富实用价值，又赏心悦目。

OPNBX的设计理念是令建筑与景观和谐统一。尽管建筑设计由其他的设计公司完成，紧密的合作保证了建筑与环境的统一，让设计理念更加有力。在VIA波塔尼项目的设计中，理念形成了价值，而保护树木就是最大的价值。

6. The 2nd courtyard that consists of swimming pool and pool deck area
7-9. Details of the swimming pool and the deck

6. 由游泳池和泳池平台区所组成的第二庭院
7～9. 游泳池和泳池平台细节图

ASNIÈRES PUBLIC PARK

阿斯涅尔社区公园

Location: Asnières sur Seine/France
Completion: 2012
Design: Espace Libre
Photography: Espace Libre
Area: 6,200sqm
项目地点：法国，阿斯涅尔
完成时间：2012年
设计师：E空间景观工作室
摄影：E空间景观工作室
面积：6,200平方米

Located in the heart of Asnières sur Seine, a plot of 6,000 m^2 on two levels becomes a meeting place for the population. As an urban staple it helps build a missing link between the different neighbourhoods, college and high school, which revolves around this space. To link the two sides of the square, a fold has achieved in the axis of the two entrances thereby making more natural path between them. Playgrounds surround the plot centre of the square, of which the top level, is reserved for children under 6 years, while the lower level is open to teens. A second square plot serves to college while acting as a gateway to the development. Moreover, a species-specific mono lavender, is planted on the edge of the high level to emphasise the link between the two entrances.

The designer strives to enhance the structural elements of the existing landscape, to make functional spaces while highlighting the social bond, to support the work done by opening up new networks and roads, to showcase strong areas and centralists, as well as developing the attractiveness of equipped sites.

这座位于塞纳河畔阿斯涅尔市中心的公园总面积6,000多平方米，是市民休闲集会的好去处。作为一个城市地标，它在公园周围不同的社区和学校之间建立起联系。为了连接广场的两侧，设计师将两个入口之间的轴线弯折起来，形成了更自然的走道。环绕着广场中央的游戏区专供6岁以下儿童嬉戏玩耍，而下层空间则供青少年使用。另一个广场为大学服务，同时也相当于整个开发的大门入口。此外，上层空间的边界栽种了一种独特的薰衣草，进一步突出了两个入口之间的联系。

设计师的目标在于：提升现存景观的结构元素，在打造功能空间的同时突出社会联系，通过建立新的网络和道路来辅助设计，展现特色区域和中央焦点，开发出景观地块的独特魅力。

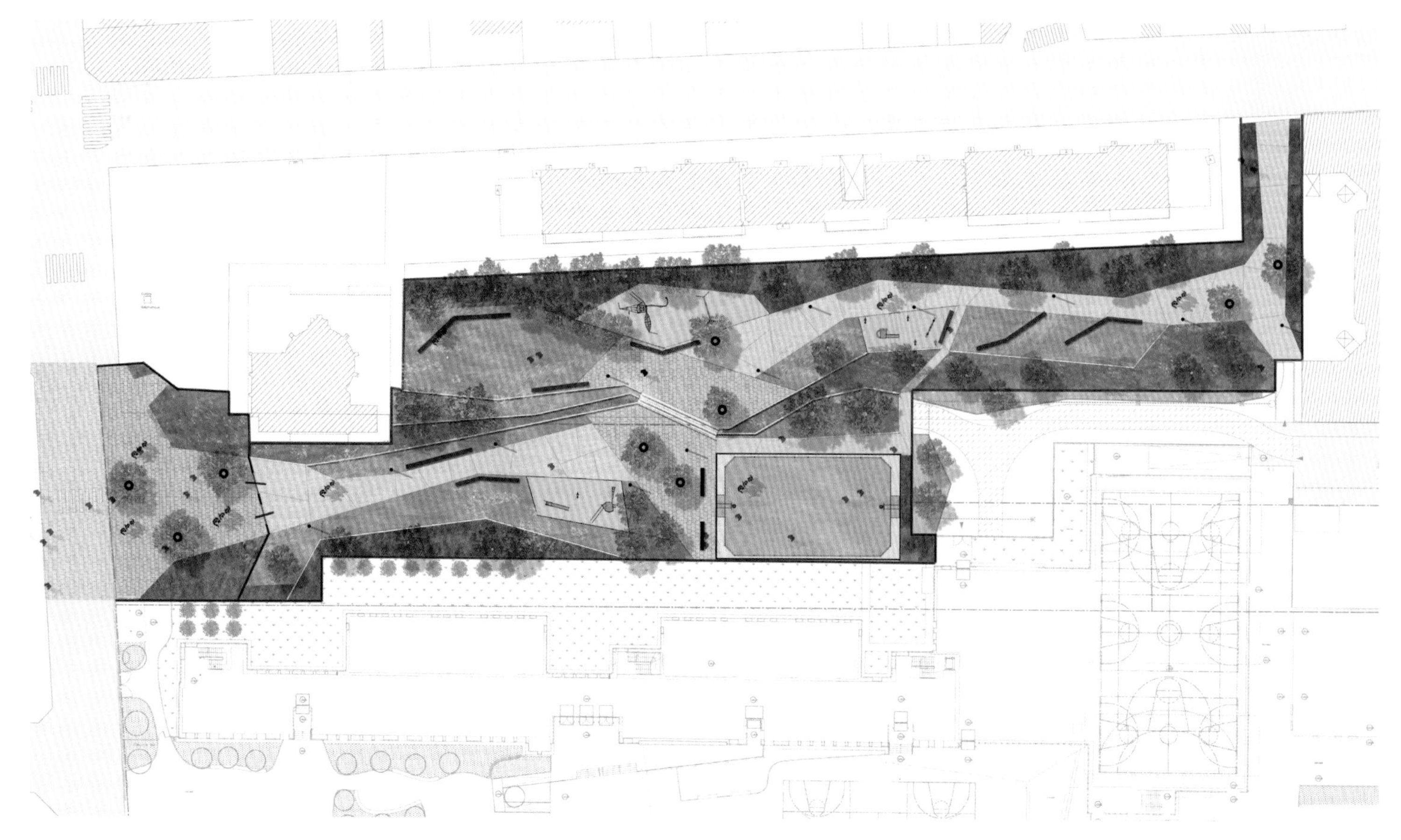

Plan drawing 平面图

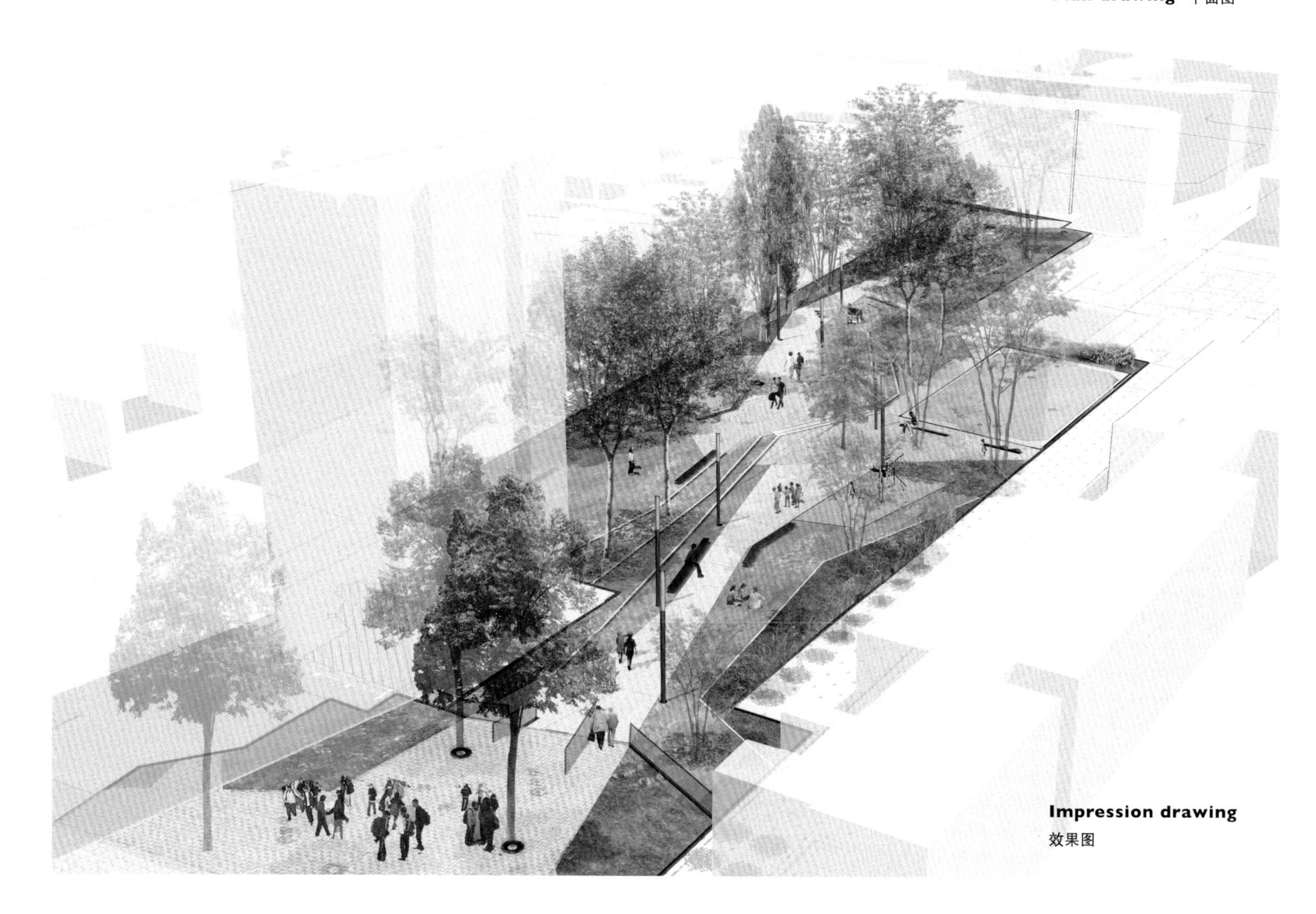

Impression drawing
效果图

1. Overview of the public park
2-3. Playgrounds surround the plot centre of the square
4-5. Paving details

1. 社区公园鸟瞰图
2、3. 环绕广场中央的游乐场
4、5. 铺装细节图

ATHLETESVILLAGE PLOT N13 & N26 LANDSCAPE – PODIUM GARDENS

奥运村N13与N26地块景观——平台花园

Location: London, UK
Completion: 2012
Design: C.F. Møller Architects
Client: ODA (Olympic Delivery Agency), LendLease
Area: 7,000sqm (podium gardens)
项目地点：英国，伦敦
完成时间：2012年
设计师：C·F·穆勒建筑事务所
委托人：伦敦奥组委
面积：7,000平方米（平台花园）

The key objective for the podium garden landscapes of the plots N13 and N26 of the Athletes Village has been to re-create lost ecology values, in a once ecologically rich region. This is in accordance with the Stratford City Masterplan and the Site Wide Biodiversity objectives for the Olympic village. The landscape design seeks to create an urban woodland, in the broadest sense of the word, with geographical and historical references to the borough of Waltham Forest and the adjacent southernmost fringes of Epping Forest.

Hence the landscape design sets the plots in the wider landscape context of the Lea Valley, Epping Forest and the open spaces of London as whole. Within the scheme the designers have achieved a rich diversity of habitats to attract wildlife to the urban setting.

The planting strategy uses a largely native palette of species encouraging biodiversity and providing a human scale to the courtyards. In order to enable planting of semi-mature woodland species, with possibility for future growth on the podium deck the design concept is based on forming a series of green hillocks. These have the dual purpose of accommodating the resident's needs for outdoor recreation in a diverse landscape and at the same time maximising the growth potential for the maturing trees. Imbedded within the lawns are ornamental and space defining beds with tall grasses and lavender planting, interconnected with tree groves and hills.

A path paved with yellow bricks threads through the courtyard creating a visual and functional connection between the built architecture and the landscape spaces. The hillocks references to the wider landscape context and simultaneously they provide identifiable orientation features to the courtyards.

1

The position of the individual woodland hills' is situated in accordance to provide shelter for the exposed eastern elevation of the courtyard whilst the large trees reinforce a sense of unity between the two plots architectural forms. The juxtaposition of consistency and contrast between landscape and built architecture creates a rich, subtle and engaging environment.

1. The landscape design seeks to create an urban woodland
2. Bird eye view

1. 景观设计试图打造一个广义上的城市森林
2. 鸟瞰图

N13 plan
N13地块平面图

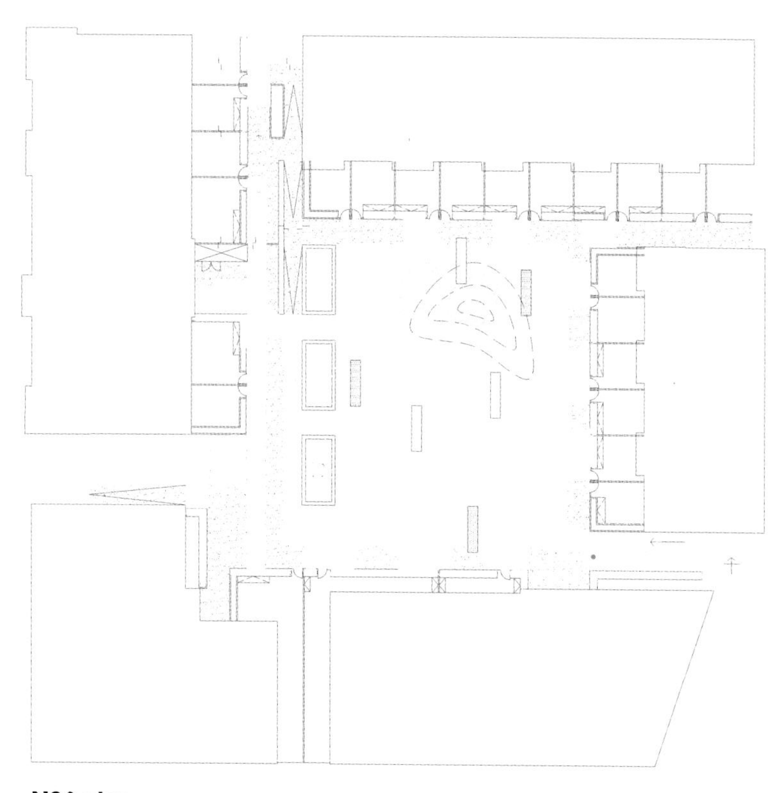

N26 plan
N26地块平面图

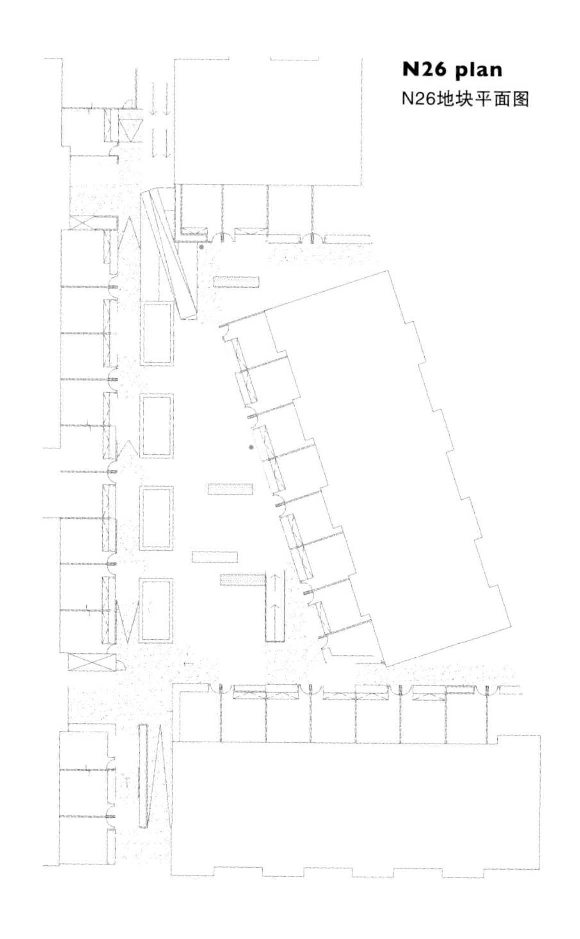

N26 plan
N26地块平面图

2

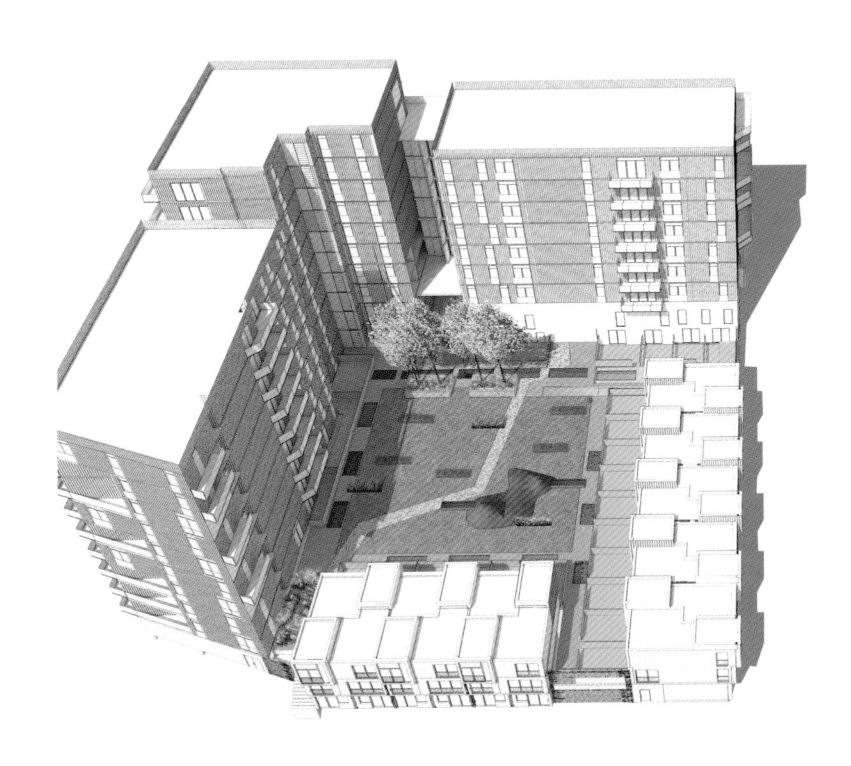

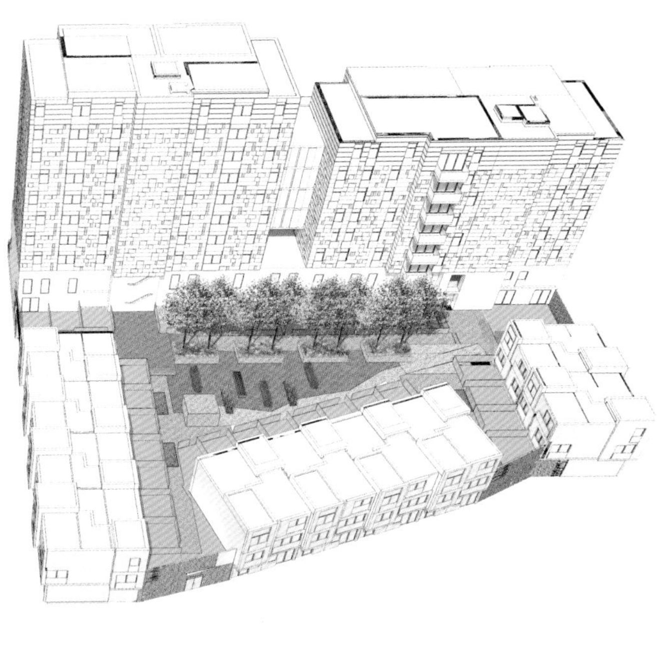

3. A series of green hillocks
4-5. A diverse landscape for outdoor recreation
6. Embedded within the lawns are ornamental and space defining beds with tall grasses and lavender planting
7. A path paved with yellow bricks threads through the courtyard

3. 绿地景观
4、5. 各种户外休闲景观
6. 草坪里嵌入了装饰花池，里面种植着高茎草和熏衣草
7. 铺装着黄色地砖的小路贯穿庭院

奥运村N13与N16地块平台花园景观的主要设计目标重新找回这个曾经拥有丰富多样的生态环境所遗失的生态价值。这与斯特拉福德市的总体规划以及奥运村的生物多样化目标相一致。景观设计试图打造一个广义上的城市森林，在地理位置和历史上参考与埃平森林最南端相邻的沃尔瑟姆弗雷斯特森林区。

因此景观设计奠定了莱亚谷、埃平森林以及伦敦的露天场所的大景观环境的基调。设计师在规划方案中实现了栖息地的多样性，帮助城市环境吸引了更多的野生动植物。

设计大量选用了本土物种来鼓励生物多样性，为庭院增加了人性尺度。为了栽种半成熟的林地物种、保证未来发展，景观设计的理念以塑造一系列绿色山丘为基础。这种设计既满足了居住者在多样化景观中对户外休闲空间的需求，又在最大程度上满足了成熟树木的生长潜力。草坪中嵌入了装饰花池和间隔花池，栽种着高茎草和薰衣草，与树林和山丘相互连接。

由黄砖铺成的小路穿过庭院，在建筑与景观之间形成视觉和功能连接。山丘的设计参考大环境，同时也为庭院提供了一定的导向特征。

独立林地山丘的位置正好为庭院东侧提供了遮蔽，高大的树木在两个地块的建筑形式之间建立起了统一感。景观与建筑的统一与对比营造出了丰富、精致、迷人的环境。

•*High-rise residential tower landscape*

高层居住区景观

THE INTERLACE
翠城新景

Location: Singapore
Completion: 2013
Landscape Designer: OMA (Concept/SD) / ICN Design International Pte Ltd., Singapore
Design Architect: OMA / Ole Scheeren
Photography: Iwan Baan
Area: 170,000sqm

项目地点：新加坡
完成时间：2013年
景观设计师：大都会建筑事务所/ICN国际景观设计公司
建筑设计师：大都会建筑事务所 / 奥雷・舍人
摄影：伊万・巴恩
面积：170,000平方米

The 170,000 m^2 development, which was completed and handed over to residents in late 2013, provides 1,040 generous residential units of varying sizes that are reasonably priced. The unusual geometry of the hexagonally stacked building blocks creates a dramatic spatial structure populated by a diverse array of activity areas.

Eight expansive courtyards and their individual landscapes are defined as the heart of the project and form distinct spatial identities. Each courtyard, spanning a distance of 60m across and extending further through the permeable interconnections, possesses a specific character and atmosphere that serves as a place-maker and spatial identifier.

The primary pedestrian route through the project leads residents from the main entrance through and to the courtyards as primary points of orientation and identification – you live in a courtyard, a space, rather than a building or an object. Pedestrian circulation is grouped and bundled according to the density of residents around each courtyard in a central "connector". A system of secondary footpaths brings residents from the connector to the private front doors of their homes.

The Interlace generates a space of collective experience within the city and reunites the desire for individual privacy with a sense of togetherness and living in a community. Social interaction is integrated with the natural environment in a synthesis of tropical nature and habitable urban space.

The notion of community life within a contemporary village is emphasised throughout the project by an extensive network of communal gardens and spaces. A variety of public amenities are interwoven into the landscape, offering numerous opportunities for social interaction and shared activities integrated with the natural environment.

A Central Square, Theatre Plaza, and Water Park occupy the more public and

1. Bird eye view
2. The unnsual geometry of the hexagonally stacked buildings creates a dramatic spatial structure

1. 鸟瞰图
2. 交错的住宅楼以六边形格局联结叠加

central courtyards and contain numerous shared amenity areas such as a clubhouse, function and games rooms; theatre, karaoke, gyms, and reading rooms; and a 50m lap pool and sun deck, family and children's pools. Surrounding courtyards such as The Hills and Bamboo Garden provide shaded outdoor play and picnic areas with lower blocks around its perimeter. The Waterfall, Lotus Pond, and Rainforest Spa complete the eight main courtyards and offer residents further choices and areas in a more contemplative environment with additional swimming pools, spa gardens, and outdoor dining.

Multiple barbeque areas, tennis and multicourts, organic garden, pet zone, and "the rock" line the perimeter of the project and offer a wide selection of communal activities for residents. A continuous loop around the site provides a 1km running track and connects the "internal" courtyards to the activities around the edge of the site.

The character of a vertical village embedded in a rich landscape of activities and nature is evident throughout the project. Elevated roof terraces and sky gardens extend outdoor space on multiple levels with views above the tree line to the surrounding courtyards, parks, sea, and city. The diversity of the various offerings and atmospheres of natural environment encourage social interaction with the freedom of choice for different gradients of privacy and sharing, contributing to the overall sense of community.

Sustainability features are incorporated throughout the project through careful environmental analysis and integration of low-impact passive energy strategies. A series of site specific environmental studies, including wind, solar, and daylight analysis, were carried out to determine intelligent strategies for the building envelope and landscape design. As a result, the project has been awarded the Universal Design Mark Platinum Award and Green Mark Gold Award from Singapore's Building and Construction Authority.

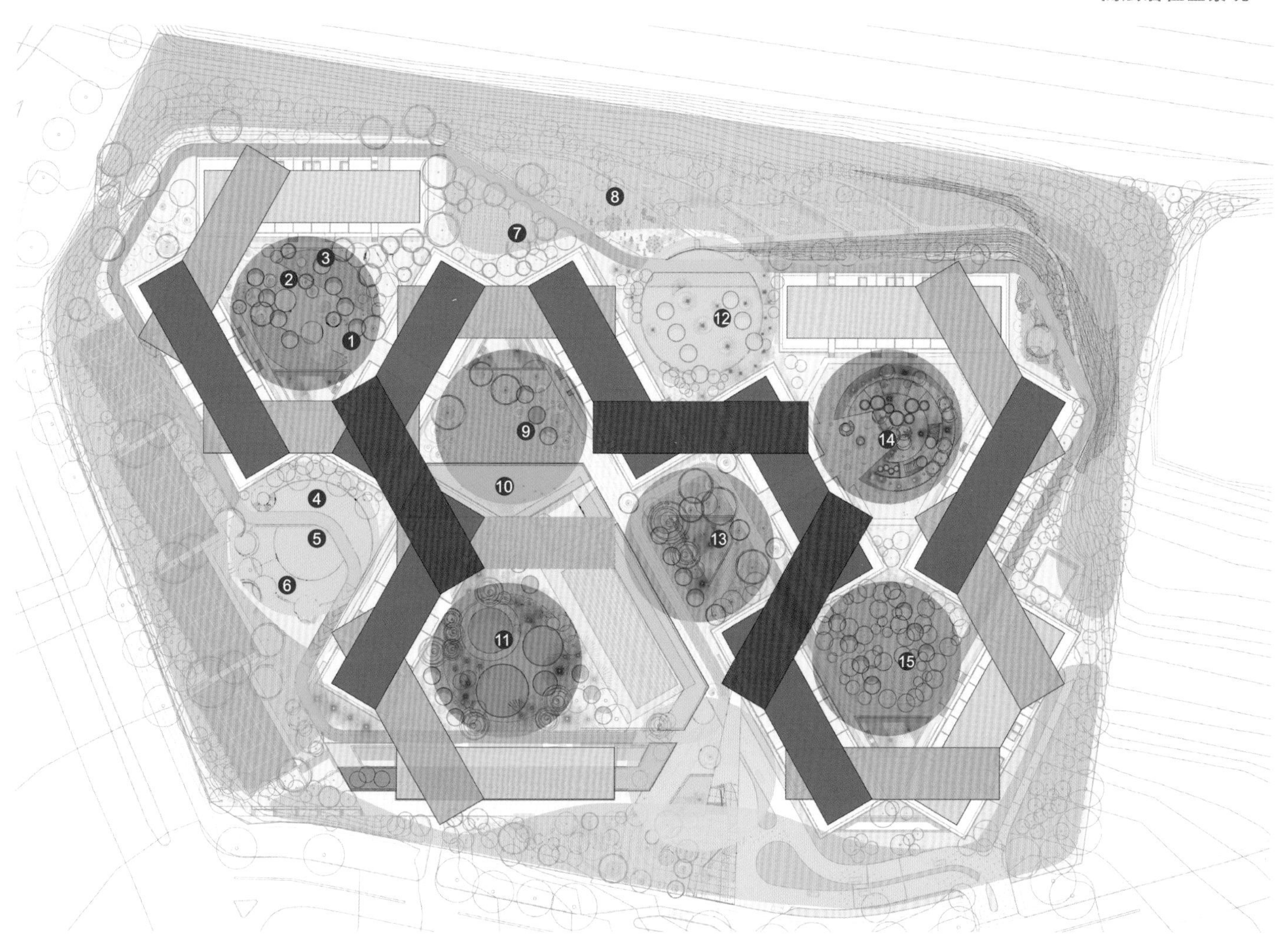

1. Bamboo garden
2. Stone garden
3. Seating areas
4. Reflection pond
5. Water fall
6. Leisure pool
7. Organic garden
8. Play zone
9. Theatre plaza
10. Gymnasium
11. Water park
12. Lotus pond
13. Central square
14. Rainforest
15. The hills

1. 绿竹花园
2. 石头花园
3. 休息区
4. 倒影池
5. 瀑布
6. 悠闲池
7. 有机花园
8. 游乐区
9. 戏剧广场
10. 健身馆
11. 水上乐园
12. 莲花池
13. 中央广场
14. 雨水森林
15. 小山

Site plan
总平面图

The roof scape
屋顶景观

这个总面积170,000平方米的开发项目于2013年末完工并交付使用，拥有1,040套不同规格的住宅单元。交错的住宅楼以六边形格局联结叠加，营造出夸张的空间结构，楼间点缀着景观各异的活动区。

8个宽敞的景观庭院及其独特的景观特征是项目的核心，形成了独特的空间特色。每个庭院的跨度可达60米，各个庭院相互渗透联系，以独特的特征和氛围起到了场所营造和空间定义的功能。

项目的主要步行路线引领着居民从主入口穿过庭院或到达庭院，让人感觉到自己生活在庭院中、空间里，而不是生活在某座建筑或某个物体里。步行交通根据住户的密度围绕着庭院集中起来。次级走道系统将居民从庭院“连接处”带到他们各自私宅的房门前。

3

翠城新景小区在城市中创造了一种集体体验空间，既满足了个体的隐私需求，又实现了一种社区的凝聚感。社交互动被融入到了具有热带风情的自然环境和宜居城市空间之中。

现代村落的社区生活概念通过丰富的公共花园和空间体现出来。各种各样的公共设施与景观交织起来，在自然环境中为社交互动和共享活动提供了丰富的机会。

中央广场、剧院广场和水上公园占据了更开放的中央庭院，包含丰富多彩的共享设施，例如：俱乐部、活动室和游戏室；剧场、卡拉OK房、健身房和阅览室；50米标准泳池及日光浴平台、家庭及儿童泳池等。环绕庭院的小山和竹园提供了清凉的户外游戏和野餐空间。其他庭院空间为瀑布、莲花池和雨林温泉，它们为居民提供了多种多样的休闲选择，配有游泳池、温泉花园、户外就餐区等功能空间。

项目外围设有多个烤肉区、网球场、有机花园、宠物区以及假山园，为居民的公共活动提供了广泛的选择空间。环绕项目场地的环形带

4

3. Dramatic spatial structure populated by a diverse array of activity areas
4. The swimming pool
5. Pedestrian circulation is grouped and bundled
6. A variety of public amenities are interwoven into the landscape
7. Night view of the community

3. 楼间点缀着景观各异的活动区
4. 游泳池
5. 步行交通根据住户的密度围绕着庭院集中起来
6. 各种各样的公共设施与景观交织起来
7. 小区夜景

提供了1千米的跑道，并且将内层的庭院与外围的活动区连接起来。

垂直村落的特征深嵌在丰富的景观活动和贯穿项目的自然环境之中。屋顶平台和空中花园将户外空间带到了各个楼层，透过树冠展现了庭院、公园、城市乃至海洋的风景。各种各样丰富的自然环境促进了居民的互动，从而提升了整个小区的凝聚感，人们可以根据需要选择各种层次的私密或共享空间。

项目的可持续特征体现在缜密的环境分析和低影响被动式能源策略的应用上。设计师对场地进行了全方位的环境研究，包括风力、太阳能和日光分析，从而决定了建筑外壳和景观设计的智能策略。最终，项目获得了新加坡建筑局颁发的通用设计标志白金奖和绿色建筑标志金奖。

MIRO
米罗住宅

Location: Singapore
Completion: 2013
Design: ONG&ONG Pte Ltd
Photography: See Chee Keong
Area: 14,191sqm

项目地点：新加坡
完成时间：2013年
设计师：王及王有限公司
摄影：施志强
面积：14,191平方米

Miro is a high-rise residential development located within close proximity to Singapore's bustling areas of Orchard Road and Little India. It maximises the site's limited land area through an effective use of space and uses lush landscaping to give its urban buildings a touch of nature.

By placing the approach to the site via the relatively quiet Keng Lee Road, the drive up to Miro feels akin to being on a private driveway with a tree-lined boulevard. The road is further accentuated by linear latticed trellises teeming with luxuriant green creepers and giving the boundary wall a more porous and natural feel. These trellises run all the way up to the entrance podium and culminate in a grand reception structure, while an inclined water feature on the ground level welcomes visitors.

.Plants on the trellis' green wall are supported by an in-built irrigation system, which also helps to clear dissolved pollutants in the water. In addition, the trellis' green wall not only enhances the façade's aesthetic appeal but also brings down the building's temperature and keeps its surroundings cool.

The first and second storeys are set aside for group interaction and activities, with the latter serving as a spa haven and an extension of one's living space. This idea of an extended living space can also be seen in the landscaped terraces on the 3rd, 9th, 13th, 17th, 21st, 25th and 29th storeys. Each sky terrace has an integrated pantry for residents to dine in the garden pavilions, which are filled with a variety of plants that support a community of butterflies and birds.

The melding of city life with natural, green elements makes Miro a resort-like haven in the heart of Singapore.

1

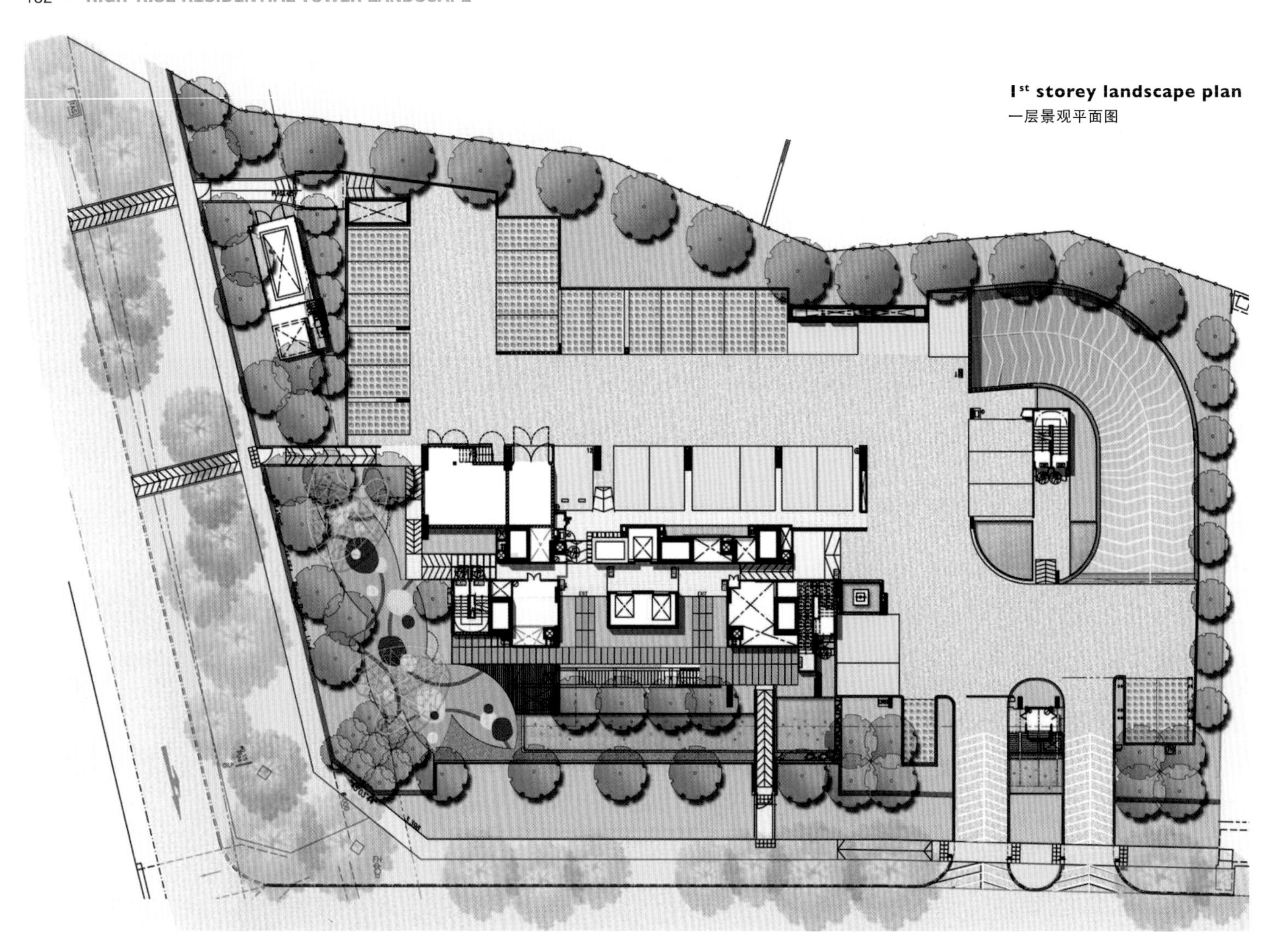

1st storey landscape plan
一层景观平面图

米罗住宅是一个高层住宅项目，紧邻新加坡最繁华的乌节路和小印度区。项目通过高效的空间利用最大化地利用了有限的土地资源，运用丰富的景观为城市住宅带来了自然气息。

项目将正门设在相对安静的庆利路上，入口车道更像是一条两侧种满树木的私人车道。爬满绿色爬藤植物的格架进一步点缀了道路，赋予了边界墙更自然的感觉。这些格架一直延伸到小区的入口平台，形成了宏大的接待结构，而嵌入的水景景观则以友好的方式向访客致意。格架绿墙上的植物由内置灌溉系统浇灌，该系统还能清洁水中的污染物。此外，格架绿墙不仅提供了立面的美感，还能降低楼梯的温度，保持周边环境的清凉。

建筑的一二层空间被划分为集体活动空间，其中二层设有水疗中心，也是住户生活空间的延伸。这种延伸的生活空间也同样出现在3、9、13、17、21、25和29层。每个空中露台都配有专门的备餐间，供住户们在花园凉亭里用餐。花园凉亭四周种满了丰富多样的植物，吸引了许多蜂蝶和鸟类。

城市生活与自然和绿色的融合让米罗住宅变成了新加坡市中心的度假天堂。

1. The drive up to Miro feels akin to being on a private driveway with a tree-lined boulevard
2. The road is accentuated by linear latticed trellises teeming with luxuriant green creepers
3. Water feature
4. Retaining wall
5. Garden bed

1. 入口车道像是一条两侧种满树木的私人车道
2. 爬满绿色爬藤植物的格架点缀了道路
3. 水景景观
4. 挡土墙
5. 花池

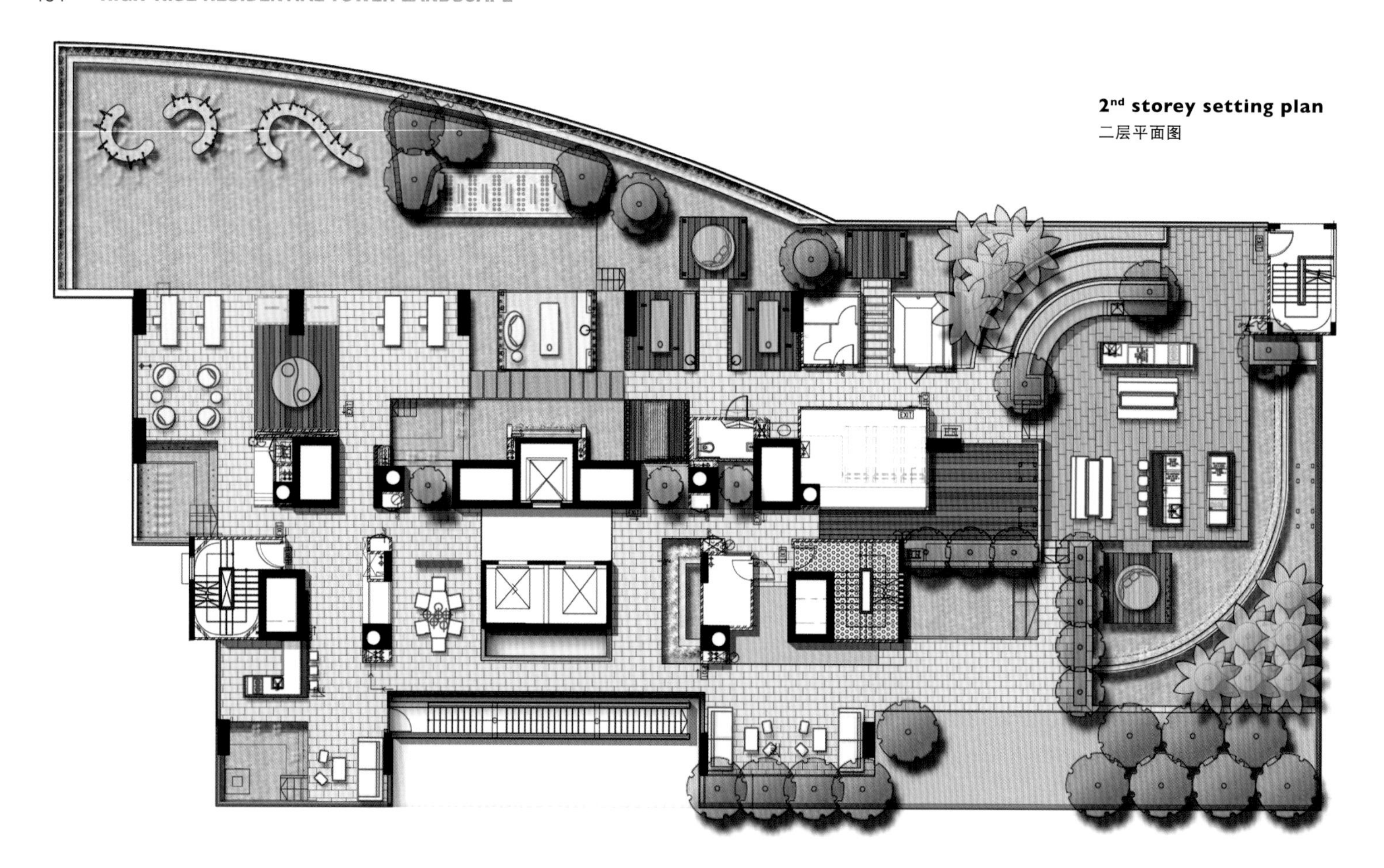

2nd storey setting plan
二层平面图

6. Garden pavilions
7. A variety of plants that support a community of butterflies and birds
8-9. An inclined water feature on the ground level welcomes visitors

6. 花园凉亭
7. 丰富多样的植物吸引了许多蜂蝶和鸟类
8、9. 嵌入的水景景观以友好的方式向访客致意

SOLEIL@SINARAN

西娜兰通道太阳公寓

Location: Singapore
Completion: 2011
Design: Tierra Design (S) Pte Ltd
Photography: Amir Sultan
Area: 12,469sqm
Award: FIABCI Prix d'Excellence International Award 2013 – Silver Winner in the Residential High Rise Category
项目地点：新加坡
完成时间：2011年
设计师：Tierra设计
摄影：阿米尔・苏尔坦
面积：12,469平方米
获奖：2013世界不动产联合会国际优秀设计奖——高层住宅类银奖

Going Urban to Suburban in Five Minutes

The Soleil condominium project lies right in the middle of one of the busiest parts of Singapore, and consequently, presented challenges from this perspective. How do you take a tenant or visitor from urban jungle to suburbia just by turning off a street?

To do this, the designers first moved the residential buildings as far as possible from the busiest road, and then lowered the car park into the basement to provide more space for design features at the entry level. This area was then covered in tropical plants and water features.

Visitors walking into the space from the busy street are immediately greeted with covered walkways that meander through the large pools and surrounding greenery, quickly putting them at ease and separating them from the hustle and bustle outside. They can walk in utter privacy and peace to their homes without prying eyes from the residents already comfortably situated above.

From the top, residents are treated to a "fifth elevation", that ensures a large undisturbed view of green and blue. The gazebo roofs are covered in curvilinear patterns of varied planting that allow observers an interesting change with the seasons. To provide variety, some plants sprout to life in the summer and others in the monsoon, making the view different at different times of the year.

The swimming areas for adults and children are separated, with smaller, shallower play areas reserved for children that are closer to the sheltered walkways and barbeque areas. Here parents can safely bask in the shelter while also looking over their children as they play.

Master plan
总平面图

1

从城区到郊区仅需5分钟

太阳公寓项目所在地是新加坡最繁华的地段之一，因为拥有上方标题所提到的困扰。如何让居民或访客从都市丛林快速转移到郊外呢?

为了实现这一目标，设计师首先让住宅楼尽量远离繁忙的街道，然后将停车场设在地下，为入口的景观设计提供了更多的空间。整个入口区域布满了热带植物和水景景观。

从闹市中走进小区的人们一下子就会被蜿蜒的小径所吸引，穿过水池和郁郁葱葱的绿化，自由自在，很快就能忘却外界的喧嚣。茂密的植物和幽静的小路让人们可以在回家的路上充分享受私密和平静，无需担心被楼上的居民窥探隐私。

在楼上，人们可以充分享受碧水蓝天和青翠植物景观。凉亭顶部覆盖着各色植物，并且会随着季节变换而形成各种有趣的变化。为了实现景观的多样性，有些植物在夏天萌芽，有些则

2

3

在雨季生长，让园区在一年四季的不同时节都有不同的景象。

成年人与儿童的游泳空间是隔开的，儿童嬉水区更浅更小，靠近绿荫长廊和烧烤区。这样一来，家长们可以一边在凉亭里放松，一边照顾游戏的孩子们。

1. Entry landscape for car park
2-3. Covered walkways that meander through the large pools and surrounding greenery

1. 停车场入口景观
2、3. 蜿蜒的小径中有水池穿过，周围被郁郁葱葱的绿色植物覆盖

4. Residents can enjoy a large undisturbed view of green and blue from the top
5. The swimming pool
6. The shelter besides the swimming pool
7-8. Lush green plants and quiet paths

4. 人们可以在楼上享受到碧水蓝天和青翠植物景观
5. 游泳池
6. 泳池旁的遮阳亭
7、8. 茂密的绿色植物和幽静的小路

Landscape plan
景观平面图

8 NAPIER ROAD

纳比雅路8号

Location: Singapore
Completion: 2012
Design: Tierra Design
Photography: Amir Sultan
Area: 6,774sqm

项目地点：新加坡
完成时间：2012年
设计师：蒂拉设计
摄影：埃米尔・苏尔坦
面积：6,774平方米

The landscape design of 8 Napier was built around accommodating and minimising the steep incline of the property. Flanked by a seven metre high wall on one side and the wall of a hospital building on the other, the design focuses on minimising views of the concrete and maximising emphasis on the greenery while moving through the property.

The generous forecourt at the entrance of the property is further enhanced by the placement of three aged Syzigium trees. These trees anchor the forecourt, creating an effortless and beautiful transition from the busy street into the property. Against a monochromatic backdrop, the rugged stone feature contrasts against the orthogonal surfaces of the pond and the property.

8 Napier is located in a very impressive part of town. Standing at 10 storeys high, comprising a total of 46 units, the architecture strives for serene simplicity as expressed by its clean, modernist lines. The indistinguishable boundary of indoor and outdoor living is elegantly composed and conceived as one. Landscape design for the scheme had to be there presentation of the unique character of the site and is in close proximity to the existing site setting, i.e. the level difference from front to rear of the site and the natural scenic qualities that the site enjoys. The design is an attempt to create a series of spatial experiences to mediate the seven metres from the entry plaza to the highest pool deck level.

The design concept derives naturally from the physical nature of the site, sloping up from south to north from Napier Road to Nassim Hill. In a series of large, gently rising terraces, beginning from the entrance courtyard porte-cochere, Napier Residences orchestrates its experiences one wonderful landscaped level at a time. Tranquil reflecting ponds and gently overflowing waters create soothing sounds, culminating at a luxurious swimming pool. The entrance courtyard is a generous space anchored by an inviting canopy which seems to levitate over a verdant green "wall" bathed in natural light. This green planting wall, together with the reflecting, overflowing pond, forms the main focus of the porte-cochere. Three large Syzygium

1. The landscape design was built around accommodating and minimising the steep incline of the property

1. 住宅的景观设计呈阶梯状展开

1

gratum trees anchor the plaza with the pond. Ten Plumeria trees with other surrounding lush green foliage, enhance the welcoming entrance plaza court experience.

2. The high wall climbing with lush plants
3. The indistinguishable boundary of indoor and outdoor living is elegantly composed and conceived as one
4. The vertical green wall

2. 爬满茂密植物的高墙
3. 室内外居住空间优雅地融为一体，密不可分
4. 垂直绿墙

3

4

5

6

7

5. The reflection pool
6. The shelter besides the pool
7. The swimming pool
8. Night view of the stairs

5. 倒影池
6. 泳池旁边的凉亭
7. 游泳池
8. 台阶夜景

纳比雅路8号住宅的景观设计呈阶梯状展开，弱化了场地的陡坡效果。景观区的一侧是7米高的围墙，另一侧则是医院大楼的外墙。设计的重点在于弱化混凝土的感觉，突出绿色植物，贯穿整个场地。

三棵老蒲桃树突出了入口处宽敞的前院。这三棵树在住宅与繁华的街道之间形成了出色的过渡。在单色背景下，高低起伏的石景与整齐划一的水池和建筑形成了巧妙的对比。

纳比雅8号住宅位于市中心的繁华路段，楼高10层，共有46套住宅单元，呈现出简约、现代的线条。室内外居住空间优雅地融为一体，密不可分。项目的景观设计呈现出场地的独特品质并且与原有的场地环境十分贴合，如场地前后的高度差、优美的自然景色等。设计利用一系列空间体验调节了从入口广场到水池平台最高点的7米高度差。

景观设计从自然环境中获得了灵感，整个项目由南至北从纳比雅路向那森山上升。庭院入口向上延伸的宽大台阶掀开了纳比雅8号住宅奇妙景观的序幕。宁静的倒影池和缓缓流淌的碧水发出舒缓的水声，最后以豪华游泳池作为顶点。入口庭院十分宽敞，以一个友好的华盖迎接着人们归来。在阳光下，整个华盖宛如飘浮在一面青翠的墙壁之上。这面绿墙与倒影池共同组成了入口的焦点。三棵高大的蒲桃树与水池共同标志着广场。10棵鸡蛋花树与四周繁茂的绿植共同提升了入口广场庭院的温馨体验。

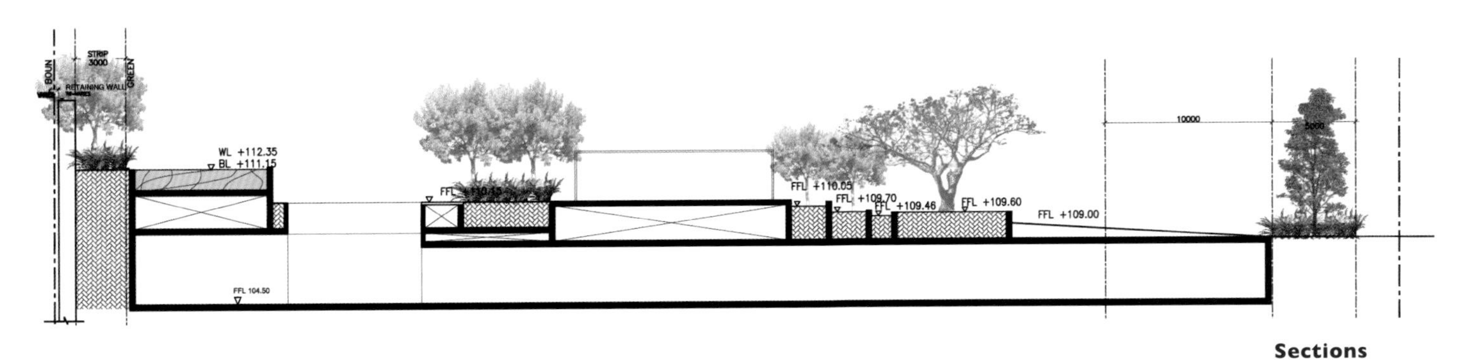

Sections

剖面图

WATERFRONT KEY

水滨丽苑

Location: Singapore
Completion: 2012
Design: ONG&ONG Pte Ltd
Photography: See Chee Keong
Area: 19,980sqm
项目地点：新加坡
完成时间：2012年
设计师：王及王有限公司
摄影：施志强
面积：19,980平方米

Located next to Singapore's Bedok Reservoir, Waterfront Key comprises of 8 blocks of 15-storey residential blocks and the site enjoys a beautiful panoramic view of the reservoir. The concept for its landscape design was to incorporate the nearby reservoir into the condominium's setting, thus making it seem like an extension from the site. Outward-facing units along Bedok Resorvoir Road will have the reservoir views while inward-looking units will be able to enjoy the central water court.

Forming the central spine of the linear water court are the lagoon and lap pools. A section of the pool deck is designed to look like a beach with a line of coconut trees leading up to where the sand gradually meets the water. This design allows to walk comfortably into the shallow water – a feature that takes into consideration the safety needs of the young, elderly and those are less confident of swimming.

Flanking this beach entry are two lushly landscaped islands with specially designed pavilion suites. These suites are available for private parties, whereby the entire island can be leased out. Each pavilion suite comprises a lounge and dining salon with a fully-fitted open kitchen. Here, one may have a soiree with friends under the stars whilst lounging by the pool with a glass of wine.

The lagoon and lap pools occupy almost three-quarters of the water court's length, culminating in the family-oriented zone where children's splash pool, aqua gym and hydrotherapy pool are clustered.

Adjacent to these pools is the clubhouse – a two-storey, linear structure that houses the gym, multi-purpose function room, steam bath as well as shower rooms. In front of the clubhouse are two tennis courts, while the clubhouse's roof terrace houses more cooking and dining facilities. A jogging circuit was also paved along the green areas of the site's periphery, and this allows avid joggers to warm up sufficiently before a run around the reservoir.

1. Outward-facing units along Bedok Resorvoir Road

1. 沿着蓄水池路的外向住宅单元享有水滨美景

WATERFRONT Key
770 772 774 776 778 & 780 BEDOK RESERVOIR ROAD

Master plan
总平面图

2. Broad-leaf, deciduous trees were also planted across the vicinity for shading purposes
3. Lush green plants growing along the path

2. 随处可见的阔叶乔木同样起到了遮阳的作用
3. 小路两旁生长的茂密的绿色植物

Interesting spatial experiences have been created through the harmonious and seamless interphase between the verdant landscape, hardscape and buildings. They have been designed to provide a conducive environment as well as opportunities for chance encounters to forge friendship in addition to family bonding.

Natural elements have been incorporated within the premises for both pragmatic and aesthetic purposes. For instance, creepers, instead of synthetic materials, were used as natural sun-shading screens for the walkways. Broad-leaf, deciduous trees were also planted across the vicinity for shading purposes. Other green measures adopted were the use of energy-efficient LED lights in water features and walls.

Waterfront Key, with its lush landscape, captures the essence of a sleepy fishing village found in the early days, while modern-day comforts necessary for contemporary living are well provided within the premise.

4. A conducive environment for residents
5. The harmonious and seamless interphase between the verdant landscape

4. 良好的社区环境
5. 绿植景观和硬景观相互渗透，和谐统一

6

7

8

9

10

11

新加坡水滨丽苑坐落在勿洛蓄水池岸边，由8座15层高的住宅楼组成，尽享蓄水池的壮丽美景。水滨丽苑的景观设计概念将蓄水池与公寓楼融合起来，使蓄水池看起来像是住宅区的延伸。沿着蓄水池路的外向住宅单元将享有水滨美景，而内向住宅单元则可享受中央水庭的景观设计。

作为中央轴线的直线型水庭由**潟**湖和游泳池组成。泳池平台的一部分被设计成沙滩的感觉：在沙子与水相接的边界上种植着一排椰子树。这一设计让人们可以舒服地踏进浅水区——浅水区的设计充分考虑了儿童、老人及不会游泳的人的安全需求。

沙滩两侧是郁郁葱葱的景观岛，岛上设有特别设计的凉亭套房。这些套房可以举办私人派对，因为整个小岛可以整体出租。每个凉亭套房都有休息室、餐厅以及设备齐全的开放式厨房组成。人们可以在星辰下与好友在池边谈天说地，浅酌一杯美酒。

潟湖和游泳池几乎占据了水庭长度的四分之三，以家庭泳池区为终点。家庭泳池区汇聚了

儿童嬉水池、水上健身区以及水疗池。

紧邻泳池的是两层楼高的俱乐部，内设健身房、多功能活动室、蒸汽浴以及淋浴室。健身房前方是两个网球场，而俱乐部的屋顶平台上则设置着更多的烹饪和餐饮设施。慢跑环道沿着项目外围的绿化区铺设，为慢跑者在环蓄水池跑步之前提供了充分的热身空间。

绿植景观、硬景观和建筑相互渗透，和谐统一，提供了丰富多彩的空间体验。它们共同组成了良好的社区环境，让人们有机会建立更密切的人际关系。

项目外围的自然元素兼具美观性和实用性。例如，爬山虎为人行道提供了天然的遮阳屏；随处可见的阔叶乔木同样起到了遮阳的作用。此外，项目还采用了节能LED灯为水景和围墙进行照明。

水滨丽苑以其丰富的景观重现了世外桃源的景象，同时又为人们提供了现代生活所必须的便利设施，是一处优雅精致的现代居所。

6. The open plaza
7-8. Playground
9. Pavilion suites
10. The featured pavement on the pool
11. The pool deck
12. Lushly landscaped islands
13. The water feature in the pool

6. 开放宽敞的广场
7、8. 儿童游乐区
9. 凉亭套房
10. 泳池的特色铺装
11. 泳池平台
12. 茂密的景观岛
13. 水景设施

39 BY SANSIRI CONDOMINIUM

新西里39公寓

Location: Bangkok, Thailand
Design: Shma Company Limited
Photography: Wison Tungthanya
Area: 2,400sqm

项目地点：泰国，曼谷
设计师：Shma景观设计公司
摄影：威森·唐桑亚
面积：2,400平方米

39 by Sansiri is a high-end condominium development in one of the busiest districts of Bangkok. Being in such prime location, building footprint occupies most of the site to maximise residential area. Outdoor space becomes an integral part of the building itself with parts of the space being covered by roof. The concept is to define this intermediate space between exterior and interior.

In order to construct a tall building in limited area, architecture is built up on wide but thin columns to maximise functioning space. Likewise, landscape is designed to perform well with architectural language, by an extrusion of outdoor space that carves into a lobby interior visually connected through transparent glass wall. And the pool on level 9, elongates above landscape and surpasses beyond building façade, which seems to be floating in the sky.

As well as an interior language was adapted into landscape design, trimmed bush and ground covers were arranged in dynamic dimension to form spatial space; concrete structures become walls and seats, creating isolate sanctuary that is set apart by green lawn and water platforms. A threshold from open yard to lobby is a long corridor with water feature on both side, accommodating an ambience effect which soothes residents who pass by. Continuous glass wall of the lobby that pushed further in is not only replenishing the lack of outdoor area but also visually connecting an experience when moving through.

On level 9, a pool extends toward the east where it is an entrance to building. An actual pool structure is cantilevered to provide lap length and extension to city views. Water body stands in the middle between sun beds on wood deck and private pavilions which place isolated on water. By facing east, architecture benefits landscape through having proper sunlight flooded the entire floor. In addition, with open wall design, the area will always receive cool breeze from the south which is essential to Thailand. Wind flows across the pavilions onto water, its surface conceives vibration then remedial treatment occurs.

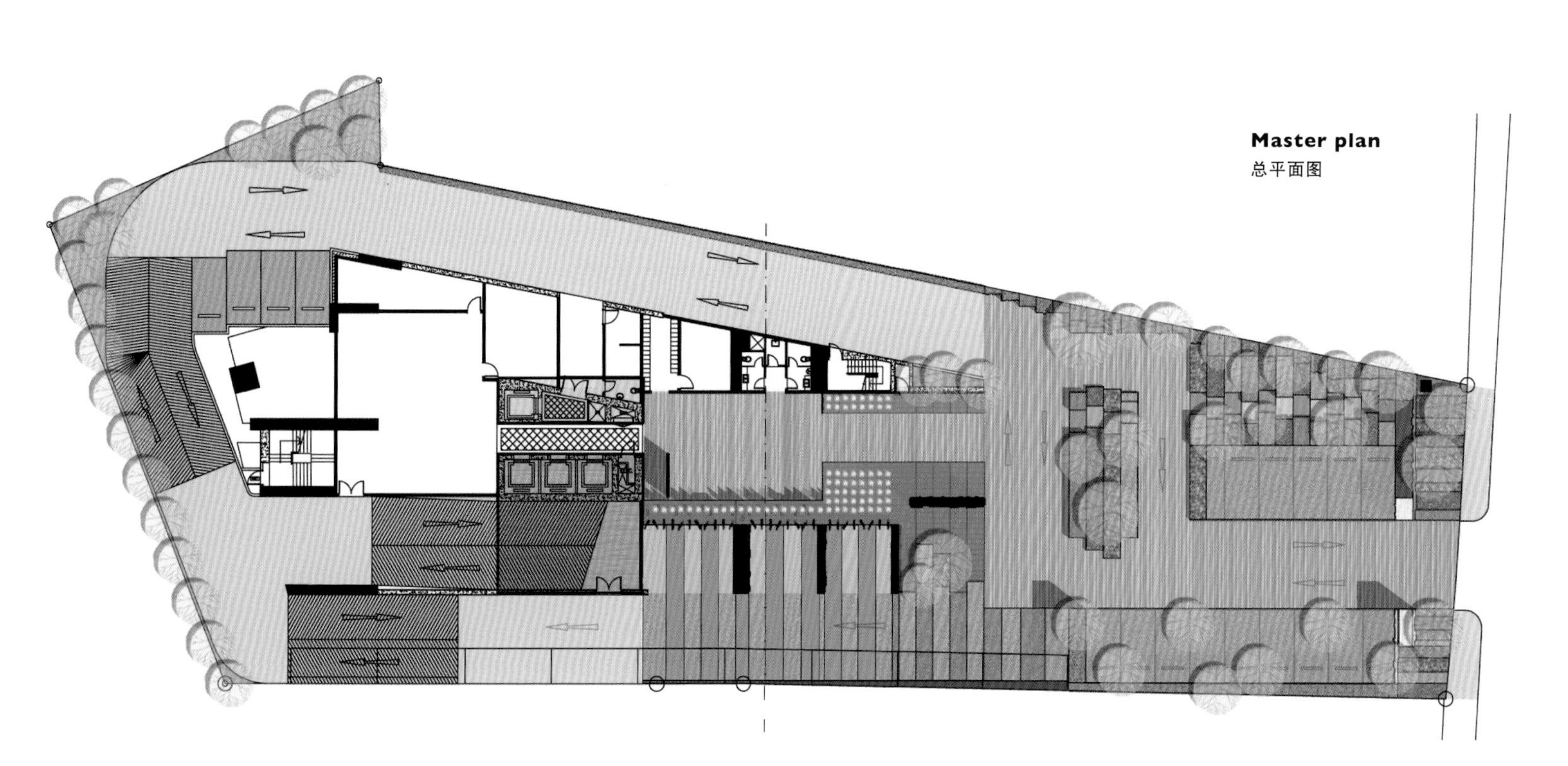

Master plan
总平面图

1. landscape is designed to performing well with architectural language
1. Trimmed bush and ground covers were arranged in dynamic dimension to form spatial space
2. Landscape details are also reflected upon finished materials and pavements
3. The vertical green wall
4. Trimmed bush and trees
5. Concrete structures become walls and seats

1. 景观设计与建筑语言相匹配
2. 修剪整齐的灌木和地被植物以动感的形态塑造了空间
3. 景观设计与建筑语言相匹配
4. 垂直绿墙
5. 修剪整齐的灌木
6. 混凝土结构形成了墙壁和座椅

5

6

For sculpturous feature, the water pavilions are finished by porous pattern woods and green hedge around it to form a personal island. Each of it stands alone on water surface that sunken at angle depth. Resting is more dramatic when each pavilion contains a private Jacuzzi that connects from adjacent pool, accessible from water deck or pavilion seats.

At night when all the lights up, central stair core that sits at the pool edge has turned into a tall lantern. Warm light suffuse through its translucent wall and the pool reflect glares along with scattered dot lights installed at pool floor. Therefore, this cantilever pool becomes a star ocean soothing all residents above them.

Furthermore, landscape details are also reflected upon finished materials and pavements. All tiles align accordingly to a direction where circulation will flow smoothly in a space. Paving on partition is vertically aligned while abstractions of natural stone that appear in dynamic depth and width of each tile also convey organic matter into architecture. A continuum materials decision from inside to outside based on a compromising between interior and landscape designers, to enhances the space's fluidity.

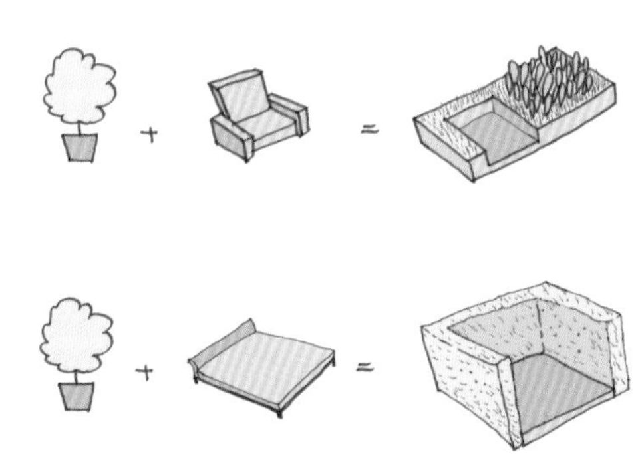

7.The pool on level 9 seems to be floating in the sky
8.Water body stands in the middle between sun beds on wood deck and private pavilions

7. 位于建筑9层的游泳池看起来就像悬浮在空中一样
8. 水体位于中央，两侧分别是日光浴平台和私人凉亭

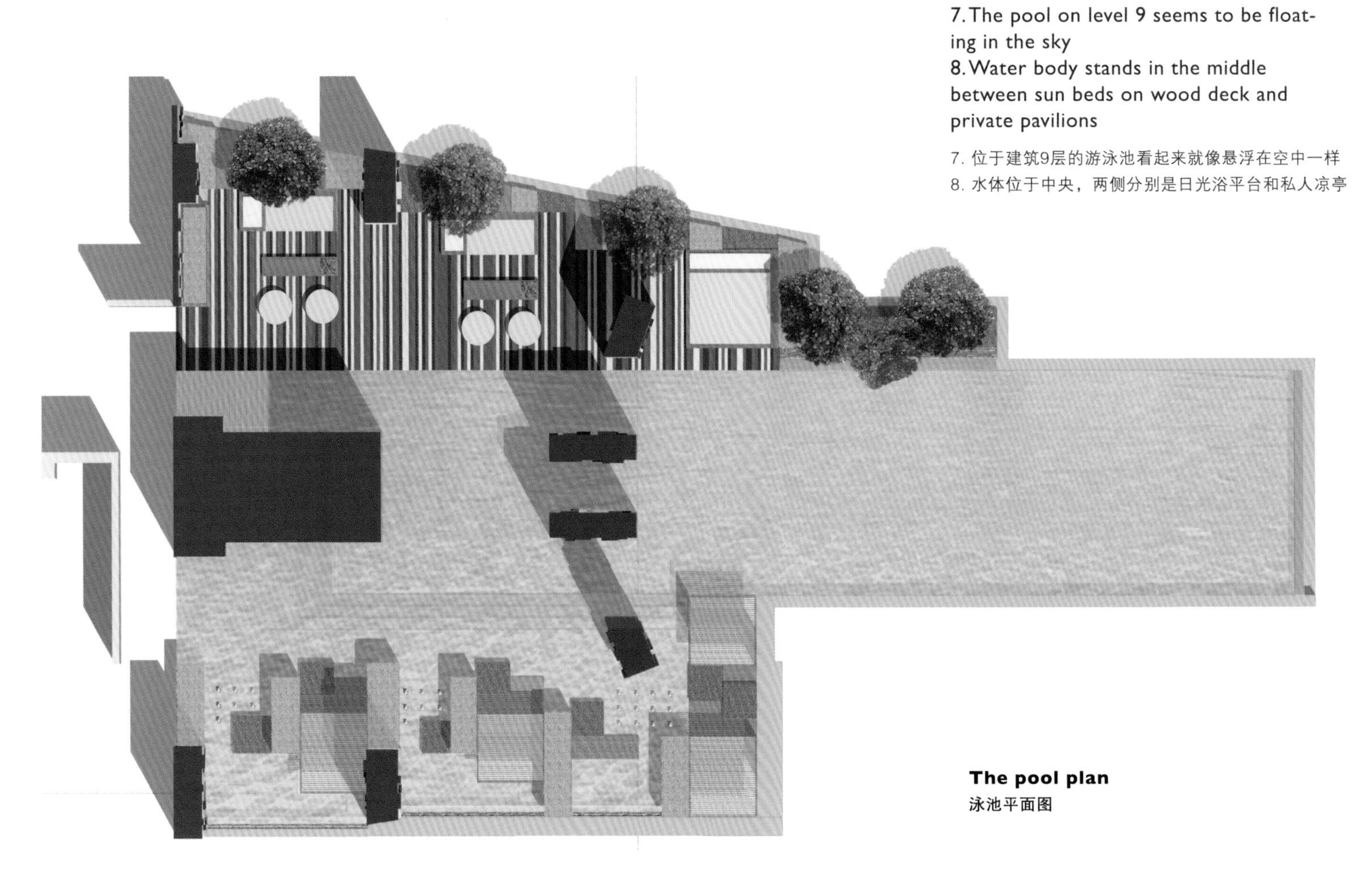

The pool plan
泳池平面图

8

新西里39是一个位于曼谷最繁华地段的高端公寓项目。由于项目位于黄金地段，寸土寸金，建筑空间占据了场地的绝大部分。户外空间成为了建筑不可分割的一部分，部分户外空间甚至位于建筑屋顶上。景观设计的整体概念为这个室内外过渡空间提供出色的设计。

为了在有限的空间内建造一座高楼，建筑通过细长的造型来增加功能空间。同样地，景观设计必须与建筑语言相匹配，户外空间切入了大厅内部，通过透明玻璃墙隔开。位于建筑9层的游泳池超出了景观和建筑的边界，看起来就像悬浮在空中一样。

室内空间同样融入了景观设计，修剪整齐的灌木和地被植物以动感的形态塑造了空间；混凝土结构形成了墙壁和座椅；草地和水景平台则打造了宁静的休闲空间。从露天庭院到大厅的入口通道是一条长长的走廊，两侧均为水景空间，让过往的住户感到安心舒适。大厅里，连续的玻璃墙不仅弥补了户外空间的缺憾，而且还提供了移动的视觉体验。

位于9层的游泳池向东侧的建筑入口方向延伸。挑出的泳池结构提供了充足的泳道长度，同时也延伸了城市景观。水体位于中央，两侧分别是日光浴平台和私人凉亭。朝南的朝向让凉亭得以享有合理的阳光和景观。此外，开放式墙壁让这一区域可以享有泰国盛行的清凉南风。微风刮过凉亭，略过水面，形成令人心情舒畅的涟漪。

至于雕塑感，滨水凉亭采用镂空的木材装饰，四周的树篱使其形成了私人小岛。凉亭独自矗立在水面上，而水池则呈斜角逐渐加深。每个凉亭都配有与相邻水池相连的按摩浴缸，可以从水上平台或凉亭的座椅直接进入。

夜晚，华灯初上，位于泳池边缘的中央楼梯井变成了一个高挑的灯笼。温馨的光线透过半透明的墙壁弥散出来，在水面上形成反光。此外，水池底部分散的点灯也为空间增色不少。整个空中泳池宛如位于楼顶的星海。

景观的细节特征还体现在装饰材料和地面铺装上。所有地砖都沿着一个方向排列，在空间里形成了流畅的交通流线。干净整洁的地砖与错落有致的天然石材铺装为建筑增添了一种有机的感觉。通过室内外景观设计师的通力合作，室内外材料的连续感让空间显得更加流畅。

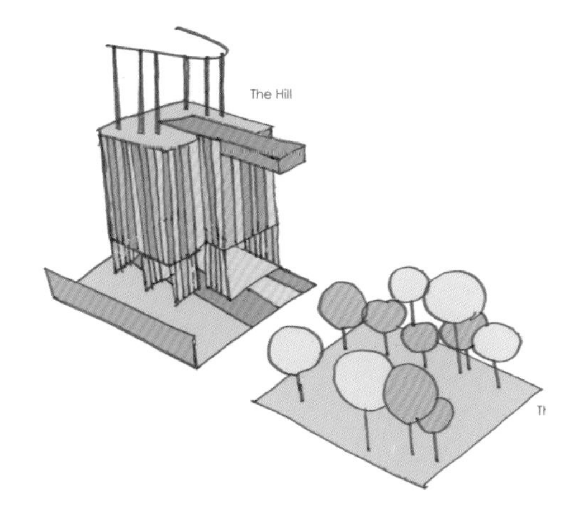

THE BASE
基本公寓

Location: Bangkok, Thailand
Completion: 2013
Design: Shma Company Limited
Photography: Wison Tungthunya, Santana Petchsuk
Area: ground floor green area: 2,363.7sqm
4th floor green area: 743.8sqm

项目地点：泰国，曼谷
完成时间：2013年
设计师：Shma景观设计公司
摄影：维森・东森雅、桑塔娜・派克苏克
面积：地面绿化区：2,363.7平方米
　　　五层绿化区：743.8平方米

People's lifestyle tends to have higher demand for individual space, while living in the densely populated district where space is limited. Condominium is one of the new. Vertical living is always the most effective solution in all cases. The extra facilities platforms are strategically added to extend the recreational space, to interlock various function together, and to form small gatherings within community. Recreational and sport facilities space are welded to each other by stripe paving patterns while selected planting palette binds landscape space as a whole. Green is also vertically linked to encourage people to use the space more productively.

As this high rise situated aligned a small street on one side and a canal bridge on the other, landscape embraces these residential units all around. Landscape specifically raised into multiple levels to serves at various functions. Therefore, the lowest ground continues from entrance toward all routes for 3 different platforms.

First route leads to an outdoor terrace, where it is extended out of the building cover, becaming a waiting platform for laundry service where parents can watch kids play on the lower ground. The second route keeps the same lowest ground height to all pockets of recreational platforms, from basketball court, kid's playground, to a more private gathering at another side of the building. The third route rises at seat height for the lowest ground, the platform is filled with green lawns and sand wash finished as some edge become seating. This last route runs along the outer edge of overall landscape that leads people to the last platform of a health exercise path with mini soccer field. All routes are completed as a full loop.

Walking around these loops makes sense when relax environment occurs. From an intense daily routine of city people, landscape not only alleviates mental health but also turns into a hangout place for friends and family. Green tall bush covers the entire concrete wall that is used for privacy. The tall bamboos alongside the vertical bridge are spanned a flowing curtain, which also provide serene sound ambience to its surroundings. Within the project's landscape is a continuous flow for an overall visual connectivity at the maximum scale.

1

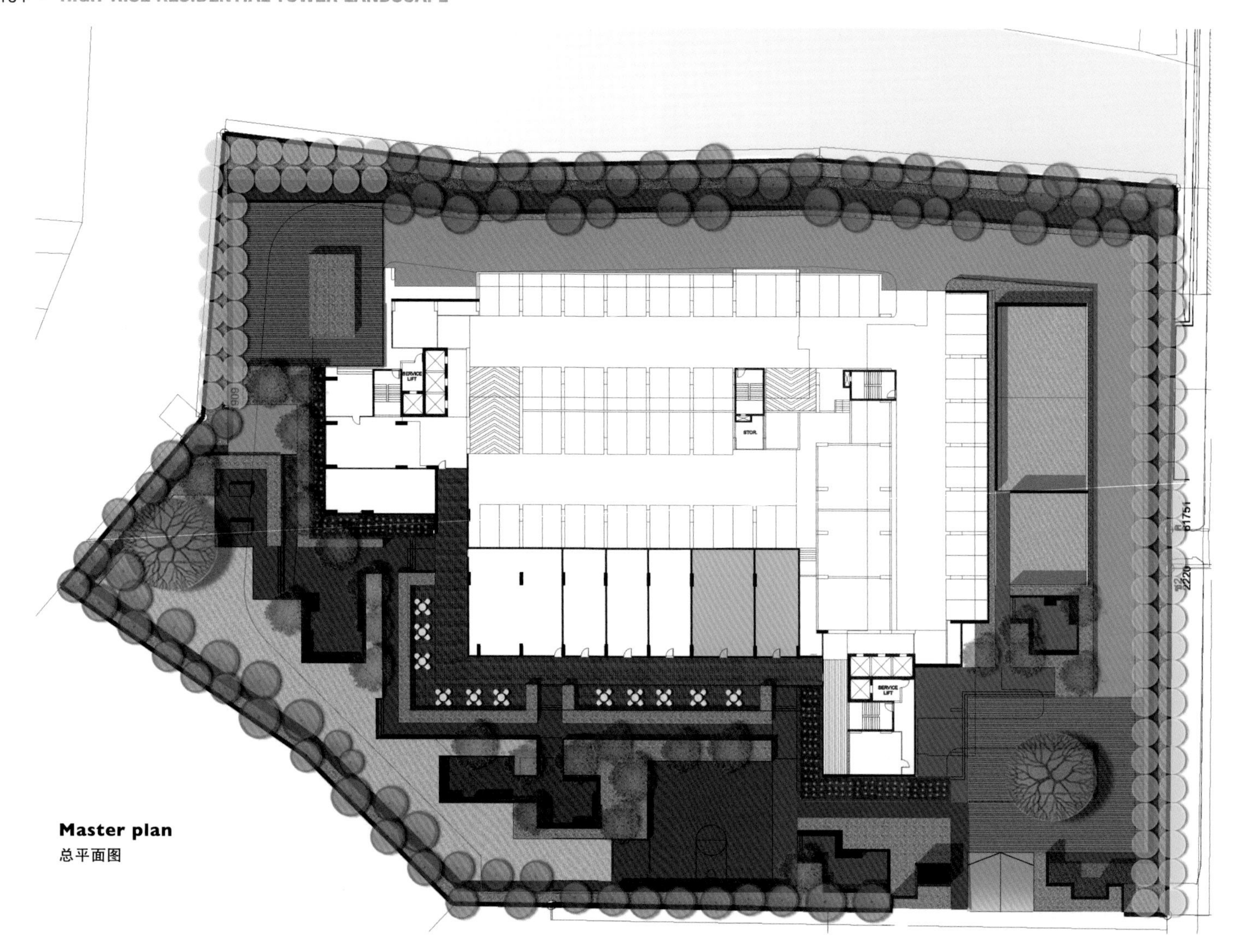

Master plan
总平面图

1. Bird eye view of the swimming pool
2. The playground

1. 游泳池鸟瞰图
2. 儿童游乐场

Consequently, the same strategy was applied to the pool and other recreational services on 4th floor. Matrix unit slightly distributes between resting and swimming area yet bonding the two different functions in an enjoyable experience. Landscape on 4th floor also is divided into 3 platforms: an Active Platform, to hide away active moments of fitness, walkway and lounge behind the semi enclosure of vertical wood fins. Pool Platform with seating cuddling in trees and bushes, placed where can be perceived from all resident levels. Flexible green platform, with only transparent glass walls that block the pool deck and cityscape.

Ultimately, the strategy of extra facility platform is used to achieve the goal. Landscape elements also are expressed in various tones of blue to create the atmosphere of youth, dynamic, energy, and mysteriousness which are out interpretation for the metropolitan character. The project not only creates a living space, it forms the community.

在空间有限的高密度社区里，人们对私人空间的要求越来越高。私人公寓是一种全新的形式，垂直居住可以说是最有效的解决方式。附加的设施平台被巧妙地添加进来，拓展了休闲空间，也将不同的功能连接起来，在社区里形成了小型集会空间。休闲和体育设施空间通过地面的条纹铺装而相互结合，而精选的植物则让景观空间融为一体。绿色空间的垂直连接也为住户提供了更加私密的空间。

项目一面临着一条小路，另一面是运河大桥，因此景观将所有住宅单元都环绕起来。多层次的景观分别对应着各种各样的功能，最低的地面从入口一直延伸到三个不同的平台。

第一条路线与露天平台相连，露台从楼体延伸出来，形成了洗衣服务的等候区，家长也可以在此照看在下面玩耍的孩子。第二条路线与休闲平台和最低的地面等高，从篮球场、儿童游乐场一直到建筑另一侧更加私密的集会空间。第三条路线与最低地面上的座椅等高，平台被绿色草坪填满，而边缘的砂洗面则形成了座椅。最后一条路线沿着所有景观的外缘延伸，引领人们前往最后一个健身平台，平台中央是迷你足球场，四周环绕着环形走道。

环绕这条走道，人们可以尽享休闲环境。在忙碌紧张的城市生活中，景观不仅能舒缓心情，还可以是亲朋好友的聚会场所。高大的绿色灌木将混凝土墙面遮挡起来，保证了空间的私密

2

3. Landscape embraces these residential units all around
4. The green plants along the path
5. Green tall bush covers the entire concrete wall that is used for privacy
6. Sand wash finished as some edge become seating
7. The platform is filled with green lawns

3. 景观将所有住宅单元都环绕起来
4. 小路旁种植的绿色植物
5. 高大的绿色灌木将混凝土墙面遮挡起来
6. 边缘的砂洗面形成的座椅
7. 平台被绿色草坪填满

5

6

7

8

9

性。两旁的竹子扩展成流畅的幕帘，营造出宁静祥和的氛围。项目的景观设计流畅连续，形成了统一的视觉效果。

泳池和其他位于五楼的休闲设施同样采用了这种设计策略。休闲和泳池区点缀着各种景观空间，它们将这两个截然不同的功能转化为愉悦的空间体验。五楼的景观同样被划分为三个平台。运动平台将健身活动、走道和休闲区隐藏在半封闭的垂直木板条后面。泳池平台配有被绿树环绕的座椅，楼上的居民对此一览无余。灵活的绿色平台仅通过透明玻璃围墙将泳池与城市景观分隔开来。

设计利用各种各样的设施平台实现了景观目标。景观元素通过各种不同的蓝色色调营造出青春活力而又略显神秘的氛围，有一种远离城市喧嚣的感觉。项目不仅塑造了生活空间，还塑造了一个社区。

8. The water feature
9. The swimming pool
10. Green platform

8. 水景景观
9. 游泳池
10. 绿色休闲平台

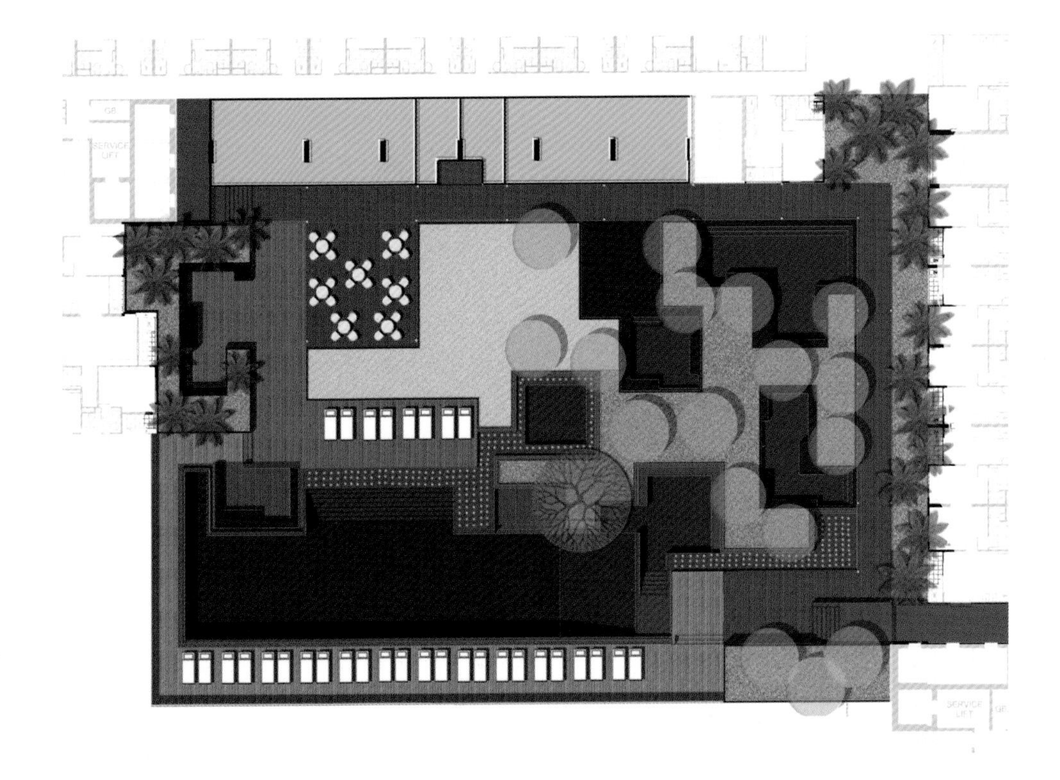

Pool plan
泳池平面图

BLOCS 77 CONDOMINIUM
77号公寓

Location: Bangkok, Thailand
Design: Shma Company Limited
Photography: Mr. Wison Tungthanya
Area: 5, 244sqm

项目地点：泰国，曼谷
设计师：Shma景观设计公司
摄影：威森·唐格坦亚
面积：5,244平方米

Blocs 77: Green Camouflage

Blocs 77 is an affordable condominium project located closely to the sky-train at one of the busiest urban area in Bangkok. The project plot is 5,244sqm and comprises of 467 residential units. The project is surrounded by many shop houses, shopping mall, and serene canal. In the front of the site, the project is facing a busy street with the traffic congestion all day; while at the back of the site, it is facing a peaceful canal and an old residential compound located on the opposite side of the canal. The rising trend of real-estate development along the sky-train route has transformed the existing low-rise houses and shop houses to the high-rise condominiums. There is the regulation of providing 6 metres fire engine route around the building and the requirement of on ground parking lots. All of which must be in a hard surface. In order to comply with the regulation and requirement, the mass of new condominium not only dominates its overall tight site and neighbours but also increases heat and glare reflecting from the building and hard surfaces around the building to the surrounding context.

Other constraints that this project is facing are the flooding during the rainy season, the high groundwater level within the site, and the limited space on the facility floor. To deal with these constraints, the landscape design approach is focused on making this project green as much as possible in order to minimise the impact of the heat and glare from the hard surfaces, raising the planting area above the existing grade level to avoid the root ball to contact with the groundwater directly, and providing sunken space to control the flooding. The designers use the concept of "tree canopy" as a metaphor of nature which helps camouflage the development with green spaces horizontally and vertically.

Design Approach

1. Turning Hard into Green

In this project, fire engine route, parking lots, and plaza areas are covered in permeable surface of turf pave and gravel. These materials help soften the

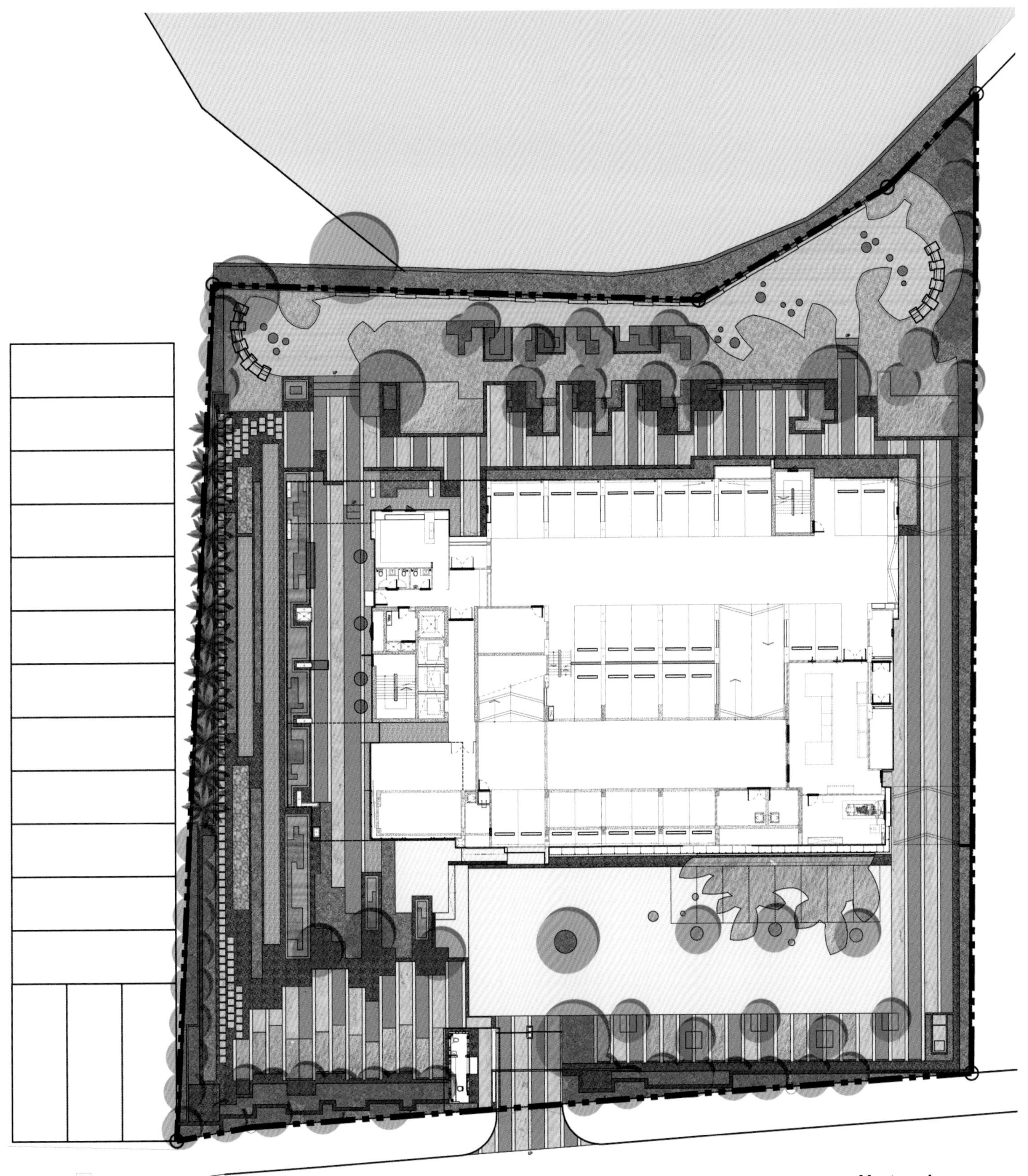

Master plan

总平面图

1

environment visually, reduce water run-off, and make these areas become more pleasant and welcome. The silhouette form of the tree canopy is used in turf pave area at the arrival court which becomes a signature of the project. At the vertical plane of the building, the landscape designers work closely with architect in order to locate the series of planters at the parking podium as well as at the residential tower. The overhanging planters at the building façade offer opportunity for the residents to get closer to nature while the planters at the parking podium help absorb carbon dioxide (CO_2) from cars. These vertical green elements not only create the unique character of the building but also reduce heat and glare from the building to its surrounding, and therefore, lead to a good environment of the city.

2. Raising above the Water Level

As the site location is adjacent to the canal and the change in urban planning from the previous permeable surface to the hard surface environment, these factors have an impact to the high groundwater level within the site which results in the difficulty in draining stormwater into the ground, the increase of water run-off, and the survival of many plant species. To solve the problem, the designers raise most of the planter areas above the ground by 450 mm to avoid root ball to contact with the groundwater directly or submerge in the water during the heavy rainfall. These elevated planters are not used as the groundwater protection but are functioned as the sculptural seating,

1. At the back of the site, the project is facing a peaceful canal

1. 项目后面朝向平静的运河

forming outdoor room for recreational purposes.

3. Edge Condition

This project is facing public areas on two sides. One is the busy road at the front and another is the tranquil canal at the back. The design approach is to maintain the continuity of the green area along both sides as much as possible. At the front edge adjacent to the street, the designers propose to set the boundary wall about 2 metres within the boundary line at some areas to provide shading green area for the public walkway. This also creates the nice visual impact from the street to the development. A simple plastered and painted wall in random pattern creates a welcome frontage along the street. At the canal side, the designers create a continuity open terrace, which embraced the canal environment. At this area, the residents can relax in the tranquil garden and can enjoy the view of the canal and the old residential compound at the opposite side. This space would also benefit the overall canalscape.

4. Waterscape

The swimming pool is built on top of the parking podium on 5th floor. The given space for the facility is not large enough to serve the large proportion of the project residents. The design approach is intended to make the space look larger while maintaining the function of the swimming pool. The designers propose to camouflage all the pool functions (i.e. pool terrace, play pool, and spa pool) under the water surface in order to extend the water surface visually. One of the design characters of this pool is the series of floating planters which sunken below the water level. These planters help create the feeling of swimming amongst swamp atmosphere.

5. Microclimate and Planting

Since the building mass dominates the major spaces of the project, leaving the narrow strips along the northeast and southwest sides. Additionally, this gigantic building resulted in the blockage of the natural ventilation that should be flowing from the canal to the front garden. To deal with this problem, the semi-outdoor terrace is introduced to connect both areas together, allowing the wind to flow more continuously and creating the channeling of vista to the canal scenery.

The concept of plant selection which is based upon the function of each specific area should match the canal ecology as well. At the front garden where the lobby and parking lots are located, the plant selection at this area must help absorb pollution effectively, provide shade, and form the space of the arrival court.

2

77号公寓：绿色伪装

77号公寓是一个位于泰国曼谷繁华地段的经济适用公寓项目。项目占地面积5,244平方米，拥有467套住宅单元，被零售店、大商场和宁静的运河所环绕。项目前方正对一条繁忙的街道，而后面则朝向平静的运河以及运河对岸的老式住宅区。越来越多的地产开发项目沿着空中电车线路而建，这造成了低层住宅和商铺向高层公寓的转变。泰国法规规定建筑四周要提供一条6米宽的防火通道，并且还要建设地面停车设施。这些设施都必须是硬质地面。为了遵守相关规定，新公寓不仅要面临紧张的场地面积和与周边建筑不和谐的高度，建筑和硬质地面散发出的热量和眩光还会给周边环境造成不良影响。

项目所面临的其他限制包括雨季的洪涝、场地的高地下水位以及有限的施工空间。为了解决这些问题，景观设计将重点放在以下几个方面：实现绿化面积最大化，以最大限度减少硬质地面所产生的热量和眩光；垫高种植区，避免植物根块与地下水的直接接触；利用下沉空间来控制洪涝。设计师利用“树冠”的自然概念，通过水平和垂直的绿化空间来“伪装”整个开发项目。

设计方案

1. 将硬景观变“绿”

在该项目中，消防通道、停车楼和广场全部覆盖着可渗透的草皮和碎石铺装。这些材料能够帮助软化环境、减少雨水径流，让这些区域变得更舒适和友好。树冠的剪影造型体现在项目的标志——到达庭院的草皮铺装区上。在建筑的垂直立面上，景观设计师与建筑师共同合作，在停车楼和住宅楼上设置了一系列花槽。从建筑立面突出了花槽既让居民更贴近自然，又能吸收汽车所排放的二氧化碳。这些垂直绿色元素不仅为建筑带来了独特的特征，还减少了建筑对周边环境释放的热量和眩光，从而对城市环境做出了积极的贡献。

2. 抬高至地下水位之上

由于项目紧邻运河，并且城市规划的硬质地面环境是由从前的渗透性地面改造而成的，二者都造成了项目场地的地下水位过高，从而不利于雨水向地下的排放，增加了地面径流，扼制了许多植物的生存生长。为了解决这一问题，设计师将大多数种植区都抬高于地面450毫米，避免了植物根块与地下水的直接接触，也

减少了暴雨中雨水的泛滥。这些抬高的花池不仅实现了地下水保护，还为居民提供了座椅，形成了户外休闲空间。

3. 边界环境

项目有两面朝向公共区域：其中正面朝向繁忙的街道，北面则朝向宁静的运河。设计方案力求尽量保持两侧绿化区的延续性。在临街一面，设计师利用约2米高的边界墙为公共人行道提供了阴凉的绿色。这一设计还提升了项目在街道上的美观度。简洁而造型随意的涂漆石膏墙营造了友好的入口氛围。在运河一侧，设计师打造了连续的开放平台，将运河环境包围起来。居民可以在宁静的花园中休息，享受运河以及对岸老住宅区的美景。反过来，这个区域也为运河景色增添了色彩。

4. 水景

游泳池建在停车楼的顶楼六楼之上。指定的空间不足以服务所有居民，因此设计方案决定从视觉上将其放大，同时保持游泳池的功能。设计师将所有泳池功能（泳池平台、戏水池、按摩池）都伪装在水面之下，从而在视觉上扩展了水面。游泳池的设计特色之一在于一系列下沉于水面之下的漂浮花池。这些花池能让人产生人在湿地中游泳的感觉。

5. 微环境与植物

由于建筑占据了项目的大部分空间，仅留下了东北和西南两条狭长的地带用于景观设计。此外，这座巨大的住宅楼阻碍了从运河到前院的自然通风。为了解决这一问题，设计师利用半露天平台将两个区域连接起来，既实现了流畅的通风，又引入了运河景色。

设计师根据各个区域的功能需求进行了植物的选择，同时还注重了运河生态环境的和谐统一。例如，前院设置着大厅和停车楼，因此这一区域所选择的植物必须能有效吸收污染物、提供阴凉、帮助塑造到达庭院的空间层次。

5
6
7

8

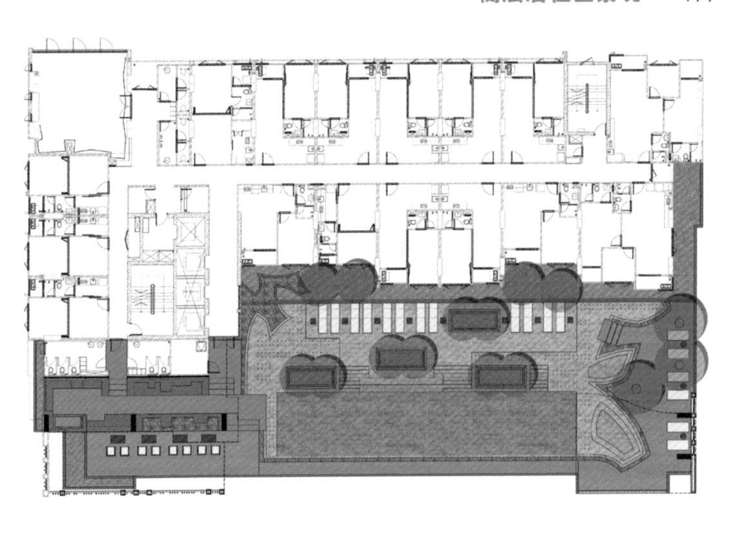

Pool plan
泳池平面图

2. Raised flower bed
3-4. Detail of the flower bed
5. The design approach is to maintain the continuity of the green area along both sides as much as possible
6. The series of planters at the vertical plane of the building
7. Shadow of the trees on the window
8-9. The swimming pool

2. 抬高的花池
3、4. 花池细节图
5. 设计方案力求尽量保持两侧绿化区的延续性
6. 建筑垂直立面的花槽
7. 窗户上的树影
8、9. 游泳池

9

BRILLIA OSHIMA KOMATSUGAWA PARK

Brillia大岛小松川公园

Location: Tokyo, Japan
Design: Yendo Associates
Photography: EARTHSCAPE
Area: 3,000sqm

项目地点：日本，东京
设计师：Yendo建筑事务所
摄影：大地景观
面积：3,000平方米

The design within the landscape design of the Brillia Oshima Komatsugawa Park creates the ability to "experience a variety of moments": from gazing over the transience of the four seasons as a part of everyday life, to feeling the rhythm of the city on your skin, to sometimes spending personal time surrounded by greenery.

The designers felt that allowing all residents to enjoy the park however they wish, rooted in their own lifestyles, was an important element of hospitality provided by the landscape of this residency, and was something that landscape design should do.

Just as an island is a stage that nurtures a plenitude of life through a diversity of locations and experiences, Brillia Oshima Komatsugawa Park is also home to a diverse sense of "time" that showcases the life here. The "Residential Island" is a stage for peoples' lives, boasting both the sense of luxury of a hotel, and a quality of lifestyle rooted in the region.

The different functions of outdoor space limit the people's behaviours. The ground of the vertical design is very simple, mainly through the ground shop outfit and greening rich visual effect.

Furniture design satisfies the human scale and material selection of enhanced sense of experience. In addition, the lamplight of setting strengthened space also makes the whole outdoor space more warmth.

2

3

Brilia大岛小松川公园的景观设计拥有让人“体验各种各样的瞬间”的魅力。从每天凝视四季的变换，到亲身体验城市的韵律，再到在绿树环绕中独享私人时间。

设计师认为居住区景观的重点是让居民以自己的生活方式享受园区景观，同时这也是景观设计最重要的任务。

正如岛屿养育着通过各种各样的栖息地和体验来滋养丰富多彩的生命一样，Brillia大岛小松川公园同样是各种展示丰富“生活瞬间”的家园。“居住岛”是人们生活的舞台，兼具酒店的奢华感和本地生活的高品质特征。

户外空间的不同功能设置限制着人们的行为。项目的地面垂直设计十分简单，主要依靠地面铺装和植被绿化来形成丰富的视觉效果。

项目的户外家具设计满足了人性化要求，通过精选的材料提升了空间体验。此外，综合的强化灯光设计让整个户外空间显得格外温馨。

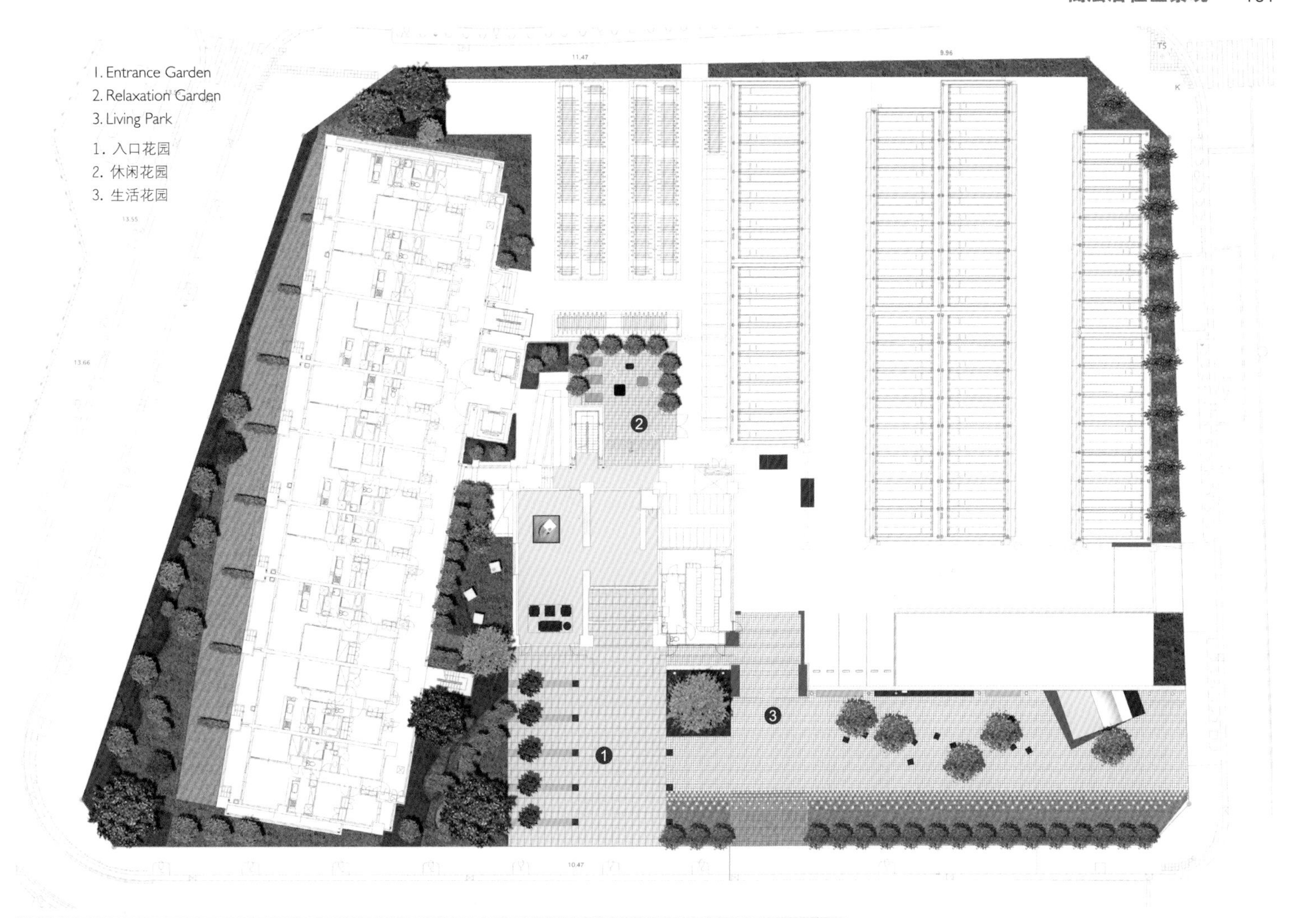

Master plan
总平面图

1. Brillia Oshima Komatsugawa Park
2. Concise pavement and the green plants
3. The seats for residents
4. Night view of the community

1. Brillia大岛小松川公园
2. 简洁的地面铺装和绿色植被
3. 座椅
4. 小区的夜景

5.The tree bed
6. Outside furniture
7.The lamplight of setting

5. 树池
6. 户外家具
7. 强化灯光设计

7

SUMMER PLACE
夏日天地

Location: Penang, Malaysia
Design: LandArt Design Sdn Bhd
Photography: LandArt Design Sdn Bhd
Area: 25,293sqm
Award: Malaysia Landscape Architecture (ILAM) Awards 2011

项目地点：马来西亚，槟城
设计师：大地艺术设计公司
摄影：大地艺术设计公司
面积：25,293平方米
获奖：2011马来西亚景观建筑奖

Sitting on 6.25 acres, off Jelutong Expressway, Summer Place is a waterfront condominium located at the northern end of the newly developed, residential area of Bandar Sri Pinang in Georgetown, Penang. A panoramic view of the sea including the Penang Bridge, the promenade and easy access to town and the bridge, helped to project the Summer Place as a seaside holiday retreat. It consists of three towers which houses more than 500 units, the upper medium cost condominium provides a full range of amenities and facilities for the pleasure and convenience of its residents. Even though this scheme comes equipped with an automation system, in order to fully complement the condo facilities, the landscape design has to be the forefront element. At the end of the day, the residents will be benefit from this comprehensive package which comes with an alluring environment.

The whole idea is about creating a bold, contemporary and vibrant garden to suit the needs of a high density, vertical living community. Besides simplicity in the appearance of the landscape, the innovative design will offset the image of this formerly, reclaimed land. This geographically sea fronting advantage should be taken into consideration when designing and transforming the landscape features. A crucial element that needs to be weaved into this concept is functionality. This is about how the amenities featured in the landscape design are able to benefit this particular community, allowing them to interact within the underlying concept of inclusion and unity in sustaining this waterfront community.

This is a contemporary, tropical garden which lies within a shrouded sanctuary. The architectural planning creates a concentrated landscaped area which comprised three tower blocks. Formerly, a dumping site, 1.2 metres of soil was removed and replaced with new topsoil in order to overcome the acidic soil conditions. As this project is about creating a holiday seaside retreat with summer activities, the water elements has been adopted as the main theme to complement the sea view context.

The centralised contemporary garden forms a gigantic circle on the plan view to

Master plan
总平面图

1

break the rigidity of the designated landscaped area. This circular shape is intended to tangibly portray a sense of inclusion, unity and togetherness in the overall design. Meanwhile, access to the blocks is highlighted with parallel alternate colours. Each lining feature paving traversed from each lift lobby, where 1.5 metre wide walk ways frame the circular garden. This pattern extends to the softscape as well as the Ophiopogon jaburan flanked by its alternate, variegated species.

This subtly raised one metre high pool deck elevated the view from the pool outwards to the sea shore. The elevated garden consists of the following amenities: main infinity pool, wading pool, timber deck cabanas, pavilion with indoor gym and Jacuzzis. Again, the parallel line pattern has been fabricated, vertically with stainless steel standing together with orangy broken mosaic water spout panels. The water element persists with water flowing to the edge of the 40 metre long, rippling infinity pool.

1. Bird view of this waterfront condominium

1. 海滨公寓项目鸟瞰图

2

3

4

2. The appearance of the landscape is simple
3-4. The featured pavement
5. The tropical garden lies within a shrouded sanctuary

2. 景观设施外观简单
3、4. 具有特色的铺装
5.坐落于一片绿树环绕的热带花园

夏日天地总占地面积约2.5公顷，靠近节路顿高速，是一个海滨公寓项目，位于槟城新开发的班达尔斯里槟榔住宅区。项目享有包含槟城大桥的辽阔海景，与市区和大桥之间都有便利的交通，是一个独特的滨海度假村。夏日天地共有三座公寓楼，提供500余户住宅，为居民提供了全套的娱乐和便利设施。项目配有自动化系统，全面配套公寓设施。景观设计师必要的前景元素。当忙碌的一天结束后，居民将在迷人的环境中尽情享受全方位的服务。

项目的整体想法是打造一个前卫、现代且生机勃勃的花园，以满足这个高密度处置社区的需求。景观设施外观简单，其设计富有创意，与这块新生的土地的形象十分相符。景观的设计和改造应当充分考虑到地块的临海优势。设计中一个重要的元素是功能性，其目标是让嵌入景观设计的设施服务于特定的社区，让居民在这个海滨社区的凝聚感和集体感中实现相互交流。

这个时尚的热带花园坐落在一片绿树环绕的世外桃源里。建筑规划创造了一个集中的景观区域。由于场地之前是一块废弃的地块，项目替换了1.2米厚的表层土来克服酸性土壤条件。项目的目标是打造海滨夏日避暑胜地，为了配合海景环境，水元素被作为整个项目的主题。

中央现代花园形成了一个巨大的圆形，打破了景观区域刻板的造型。这个圆形在整体设计中刻意营造了一种凝聚力和团结感。同时，通往各个楼的通道以交替的平行色彩来突出。特色线条铺装贯穿了各个电梯大厅，两侧1.5米宽的步行道将圆形花园环绕起来。这种形式延伸到了软景观和道路两侧的沿阶草中。

1米的高度将泳池的视野提升到了海岸之上。上升花园中包含以下设施：无边界主泳池、浅水池、观景木台更衣室、带有室内健身房和按摩浴缸的场馆。设计同样选取了平行线造型：矗立的不锈钢立柱之间设置着成为镶嵌式喷水板。水流向40米长的无边界泳池的边缘汩汩流动。

6

7

8

9

10

6. Bold, contemporary and vibrant residential garden
7. Main infinity pool
8. The water element persists with water flowing to the edge of the 40 metre long, rippling infinity pool
9-10. The playground

6. 前卫、现代且生机勃勃的住区花园
7. 无边界主泳池
8. 水流向40米长的无边界泳池的边缘汩汩流动
9、10. 儿童游乐场

XIN TIAN 40°N RESIDENTIAL DEVELOPMENT

北京北纬40° 住宅开发

Location: Beijing, China
Completion: 2012
Design: HASSELL
Photography: Yang Qitao
Area: 256,000sqm

项目地点：中国，北京
完成时间：2012年
设计师：HASSELL国际设计咨询公司
摄影：杨祺涛
面积：256,000平方米

40°N is a 13.8ha residential development located in the Chaoyang District of Beijing. HASSELL's design included the landscape design of the residential area as well as an 11.8ha adjacent public park. The name of the project, 40°N not only refers to the latitudinal location of the site, but also refers to the metaphorical relationship between space and time. Just as movement through space can bring about a diverse range of experiences, the landscape can convey a variety of impressions through its colour, shape and form. The HASSELL scheme sought to create opportunities for different experiences in the landscape – for moments of pause and for moments of passing.

A landscape spine runs diagonally across the site, linking together 5 thematic residential gardens that are formed by the staggered arrangement of residential tower blocks. Each garden conveys a different set of experiences, to either pass through or to stay. Vertical elements such as timber trellises, feature walls, pergolas, and allees of trees act to define each space. The landscape form is a juxtaposition of square and rounded elements – created by the rectilinear form of the buildings and the elliptical form of garden islands. Water is introduced throughout the site – enhancing a rich sensory experience of sound, touch and smell.

Since the project is located in Beijing, a city suffers from water shortage and with strict restriction on water consumption; effective use of water is also a focus of the project. In the long run, this unique design is not only environment-friendly, but also saves large amount of water for the residents.

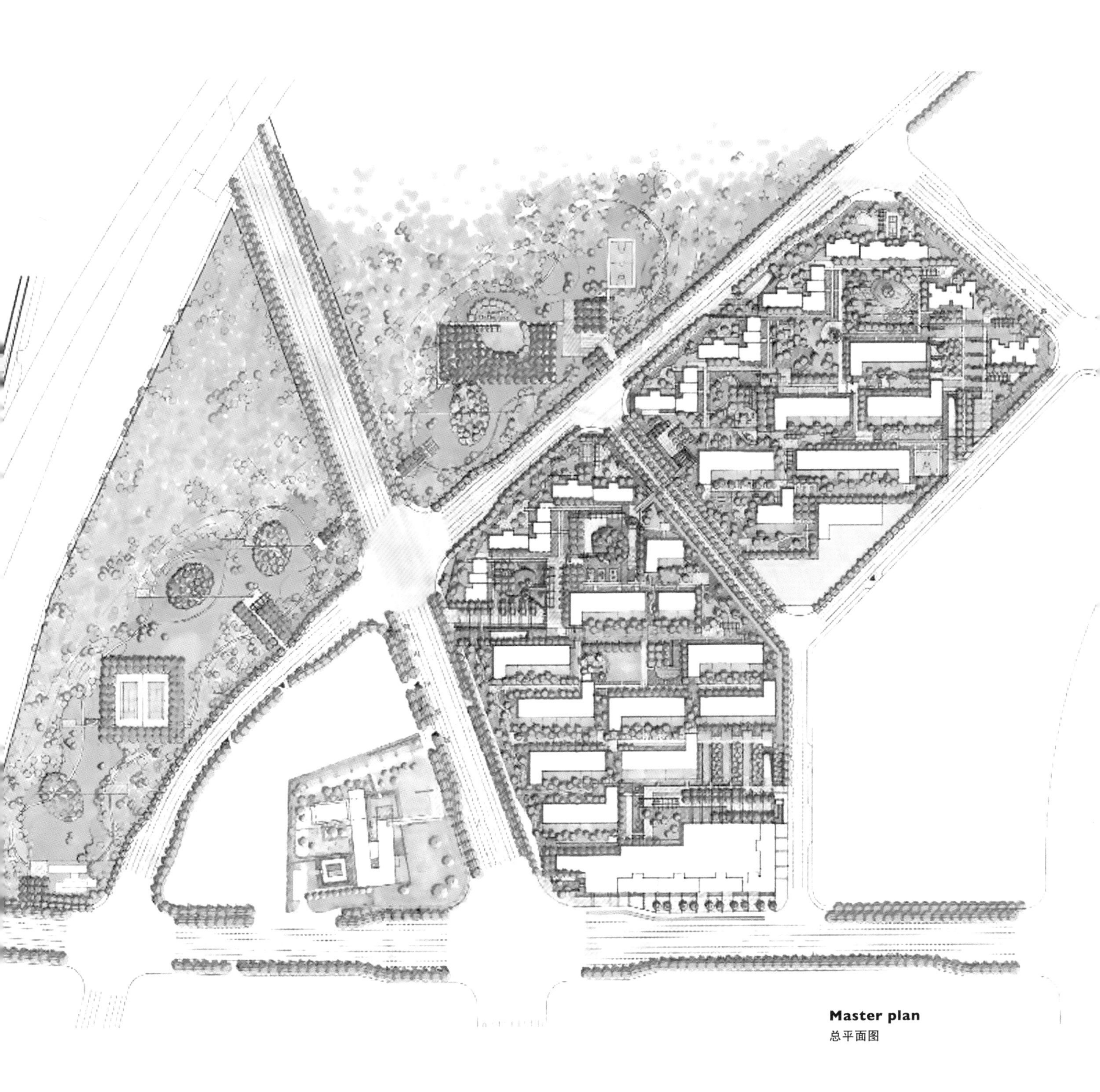

Master plan

总平面图

北京北纬40° 住宅开发位于北京市朝阳区。HASSELL受托为此13.8公顷地块以及一旁的11.8公顷“公共绿化公园”进行景观设计。项目名称“北纬40° ”不仅反映了项目所在地的纬度，还暗示着空间和时间的隐喻关系。正如在空间中移动会带来各种各样的体验，景观也能通过自身的色彩、形状和形式表达各种各样的观感。HASSELL的设计方案试图在景观中打造不同的体验，既涉及驻足停留，又涉及穿越纵横。

景观主轴贯穿了整个项目场地，由5个主题住宅花园构成。项目应用了串连景观艺术元素将这些花园连接起来。无论是匆匆走过还是驻足停留，每个花园都能向人们传达不同的感官体验。木格架、景观墙、绿廊、树列等垂直景观元素划分出各个空间的界限。景观形式体现为方形元素与圆形元素的并置，二者分别来自于建筑的直线造型和花园岛的椭圆造型。水景贯穿了整个园区，提升了丰富的听觉、触觉和嗅觉体验。

由于项目的所在地是北京，当地对用水量有所限制，因此项目的另一特点就是水源的高效使用。从长远来看，这一独特设计不仅环保，还为住户节约用水量。

1. The playground
2. The landscape form is a juxtaposition of square and rounded elements
3-4. The design seeks to create opportunities for different experiences in the landscape – for moments of pause and for moments of passing

1. 游乐场
2. 景观形式体现为方形元素与圆形元素的并置
3、4. 设计方案试图在景观中打造不同的体验，既涉及驻足停留，又涉及穿越纵横

3

4

5. Shelter	5. 凉亭
6-8. The water feature	6～8. 水景景观
9-10. The pavement	9、10. 铺装
11. The tree plaza	11. 树列广场

9

10

11

PROVINCE – PORTLAND, ZHENGZHOU

信和置业–索凌路东住宅区

Location: Zhengzhou, China
Completion: 2013
Design: Horizon & Atmosphere Landscape Co.
Photography: duo-image associates
Area: 39,473sqm
项目地点：中国，郑州
完成时间：2013年
设计师：上海翰祥景观设计咨询有限公司、瀚翔景观国际有限公司
摄影：上海柏达双影图文制作有限公司
面积：39,473平方米

Filling the Community with Youth

"Youth" stands for joy, fashion and vitality. In the landscape design of Portland Residential Project, Zhengzhou, the designers take "Youth's Vital Community" as the vision. The youth is a generation full of vigour and spirits and their community should also correspond to their characters. Therefore, the designers fill this community with joy, fashion and vitality and wish it become a unique youth community, which full of youthful feelings, passion and enthusiasm.

The combination of colours, patterns and materials provide the space with charming atmosphere. As residents walk, chat, exercise, play and enjoy views, they will indulge in pleasure and reluctant to leave.

The youth always possess a beautiful vision towards the future. Thus, with the focus on rich landscape colours and line variety, the designers also make efforts to enrich community activities and complete various supporting facilities. They even simulate activity schedule and participating groups of each area and attempt to experience the participants' feelings. The community also provides an exclusive kindergarten and thoughtful property services. Moreover, the designers further suggest the developer to transform the sales centre into an art centre which gathers children library, gym and art gallery after the housing units are sold out. They believe this measure will surely promote the community's art and cultural qualities.

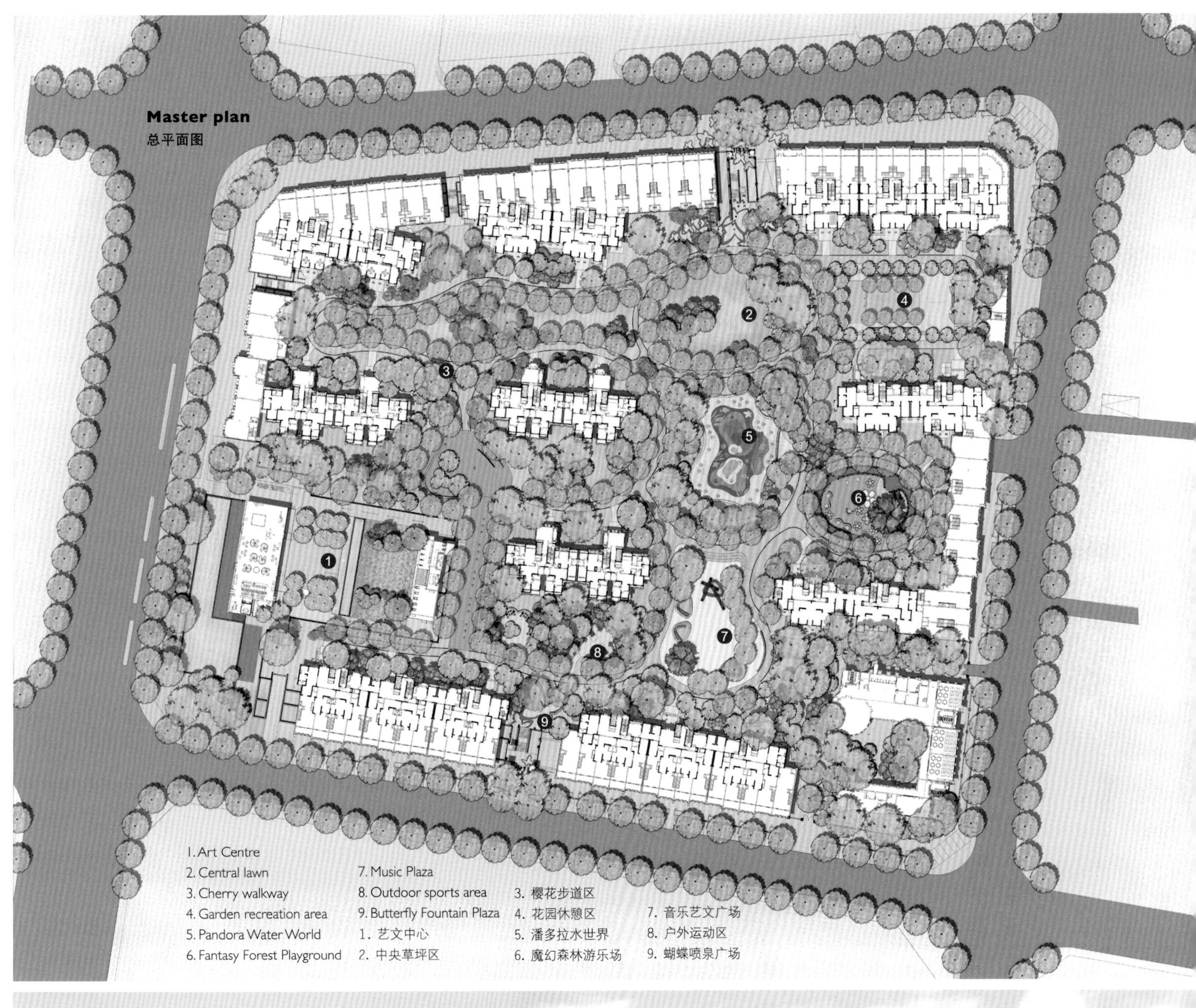
Master plan
总平面图
1. Art Centre
2. Central lawn
3. Cherry walkway
4. Garden recreation area
5. Pandora Water World
6. Fantasy Forest Playground
7. Music Plaza
8. Outdoor sports area
9. Butterfly Fountain Plaza
1. 艺文中心
2. 中央草坪区
3. 樱花步道区
4. 花园休憩区
5. 潘多拉水世界
6. 魔幻森林游乐场
7. 音乐艺文广场
8. 户外运动区
9. 蝴蝶喷泉广场

Elevation
立面图

1

2

1. The sight starts from the entry sign extended to the green lawn, tree, water feature and the architecture
2. The plaza
3. Butterfly Fountain Plaza

1. 视线由入口标示延伸至绿地、树、水景、建筑，创造空间丰富的层次感
2. 广场
3. 蝴蝶喷泉广场

让“年轻”充满社区

年轻是欢乐、时尚、活力的代名词，在进行波特兰住宅地景规划设计时，景观设计师将“年轻人的活力社区”作为这块郑州市土地的愿景。年轻人，朝气蓬勃，精力充沛的一代，他们的社区也会符合他们的性格，所以，设计师将欢乐幸福时尚活力全程灌输于这个社区中，期许其成为一个独树一帜的年轻社区。年轻的社区充满年轻人的味道，散发着青春洋溢，热情奔放的气息。

通过色彩、图案、材质的搭配，空间散发着迷人的芳香，居民在其中散步、畅谈、运动、观赏、游乐，感觉时间飞逝，乐不思蜀。

年轻人对未来生活有着更为美好的憧憬，因此，在丰富地景色彩和线条变化的同时，设计师更着力于充实社区活动，完善各类配套设施。他们甚至模拟出每个区域活动的时间及参与人群，尝试体验他们身在其中的感受。社区有专属幼儿园，也有贴心的物业服务。设计师更建议开发商将售楼处开放出来，改建为集合儿童图书馆、健身、艺术展示功能的艺文中心，提升社区的艺术文化气质。

3

4

4. Bird eye view of the Music Art Plaza
5. Olympic Plaza
6. The Music Art Plaza – sculpture that performs the dialogue between art and space

4. 音乐艺文广场鸟瞰图
5. 奥林匹克广场
6. 音乐艺文广场，雕塑上演艺术与空间的对话

7

8

7. Bird eye view of the Fantasy Forest Playground
8. Overall view of the Fantasy Forest – more extended discovery and experiences
9. Pandora Water World
10. The Fantasy Forest – creating the entertainment, sense of touch and perception into a whole

7. 魔幻森林游乐场鸟瞰
8. 魔幻森林（全部景观），更为开阔的发现和体验
9. 潘多拉水世界
10. 魔幻森林广场，创造娱乐性、触觉与知觉于一体

ORIENTAL BLESSING

杭州欣盛东方福邸

Location: Hangzhou, China
Completion: 2013
Principle designer: Li Baozhang
Associate designer: Feng Zhe, Huang Wenchang, Wu Di, Ma Wei, Yuan Li, Zhu Shuang, Zhao Qiong, Yang Zhen, Liu Tongchao, Li Dinghong
Photography: Zhang Qianxi
Area: 110,000sqm

项目地点：中国，杭州
完成时间：2013年
项目主创：李宝章
项目参与人员：冯喆、黄文昌、吴迪、马威、袁力、朱双、赵琼、杨振、刘同超、李鼎宏
摄影：张虔希
面积：110,000平方米

The project makes the most of the poetic essence of contemporary landscape in Hangzhou. The real intention of landscape designers is to emphasise "implication of image", which completes the design through materialised inner feeling, philosophic experience and association with unique images. The poetic picture drawn with a pine tree, several lush shrubs and a few flowers expresses an indescribable wonderful feeling.

The quality and artistic landscape design in Hangzhou is based on the city's extensive and profound cultural roots. On one hand, Taoist idea of "Human beings are a manifestation of the earth; the earth, a manifestation of the physical universe; the physical universe, a manifestation of Infinity; Infinity, the potential of all things" generates the principles of Chinese classic landscape design. On the other hand, the Zen idea of "Green bamboos are embodiment of truth and law, while yellow flowers are wisdom" provides theoretic support for "much in little" design of classic landscape.

By reference to local landscape concept in Hangzhou, the designers reinterpret it in contemporary methods to create a modern landscape with fresh air, warm sunshine, flowing stream and forest-like greenery. The waterfall and rockery define the entrance while exclusive loop home-entry design indicates the nobility. The surrounding sky level connects three high-rise units like a covered verandah. The view lobby juxtaposes with feature pavilion outside and the secluded paths link the exterior landscape with interior sky level. The ups and downs of rockery reinforce the depth of the garden; the rich plantings convey a natural and elegant lifestyle; the beautiful landscape provides visitors and residents carefree and content feelings.

In general, the implementation of Oriental Blessing project is excellent. Especially the landscape construction highly implements the design intention. The original landscape concept is to build a retreat for urban residents to live with nature and community harmoniously. Living with family in an environment surrounded by landscape is the ideal living state of the Chinese. The landscape design of

Master plan
总平面图

Oriental Blessing realises this result to a maximum degree. The waterfall and garden at the entrance change from winding paths to openness. The housing units are faced with delicate rockery and water features and enjoy a lush greenery view. From the elaborately stacked hills, undulating paths to rich plantings, the landscape construction constantly strives for excellence, which shows delicacy and leisure lifestyle in landscape details.

1.The most of the poetic essence of contemporary landscape in Hangzhou
2. Exclusive loop home-entry design indicates the nobility

1. 杭州现代自然山水园林所富有的诗情画意
2. 环形入户设计体现了尊崇感

该项目充分展示了杭州现代自然山水园林所富有的诗情画意。但造园者的真正目的是强调所谓“意向之象，言外之意”，即通过物化的内心情感、哲理体验和链接独特的形象联想来实现设计。园林中的一株青松、几丛翠绿、少许娇花所勾勒出的意境便使人有“所致得奇妙，心知口难言”“言有尽而意无穷”之感。

杭州园林具有如此的气质和精妙在于其博大精深的文化根源。“人法地，地法天，天法道，道法自然”的道家思想造就了古典园林上千年的基本法则，“青青翠竹，尽是法身，郁郁黄花，无非若般”的禅宗思想为古典园林的小中见大，咫尺山林提供了理论上的支持。

设计师参照杭州本土化的造园理念，以现代手法重新演绎，打造全新的杭州现代山水，兼有清新的空气、和煦的阳光、灵动的溪水、森林般的绿化。入口区掇山理水，形成瀑布和山涧，独有的环形入户设计体现了尊崇感，环抱型架空层将三个高层单元连接，如同风雨长廊。室内观景大堂与室外景亭遥相对望，通过幽幽小径将室外园景与室内架空层形成互动。起伏环绕的山系加强了园林的幽深感，丰富的种植搭配，尽显自然雅致的生活品位，优美的湖光山色，让置身其中的访客感受悠然自得的情趣。

总体而言，欣盛的施工成果非常不错，特别是景观建造，高度还原了设计意向。景观最初的构思是构建一个桃花源，让居住者们即使是身处都市，也能与自然与社区一起和谐的生活。与家人、社区一起住在山水之间，这是中国人最高的居住理想。杭州欣盛东方福邸的景观实景，最大程度的实现了这样的一种效果，入口处的瀑布流水、花园亭榭，从曲径通幽到豁然开朗，园林区楼房面山望水，满目青翠。不论是精心堆叠的环绕山系，还是起伏有致的园路，丰富的植物搭配，景观施工的工艺做到了精益求精，从细节处体现了整体景观设计的精致和一种悠然自得的生活品位。

3. The rockery at the entrance area
4. Garden pavilion
5. The waterfall at the entrance area

3. 入口区的山洞
4. 花园亭榭
5. 入口区的瀑布

4

5

6

6. The rich plantings
7-8. The secluded paths

6. 丰富的种植搭配
7、8. 幽幽小径

LAKE VIEW SETTLEMENTS, SHAOXING

景瑞绍兴曦之湖

Location: Shaoxing, China
Completion: 2014
Design: SED Landscape Architects Co.,LTD.
Photography: SED Landscape Architects Co.,LTD.
Area: 44,845sqm

项目地点：中国，绍兴
完成时间：2014年
设计师：SED新西林景观国际
摄影：SED新西林景观国际
面积：44,845平方米

The designers divide the whole project into two areas: South Area and North Area according to their programmes. North Area focuses on functionality and forms a fully functional and generous space effect. South Area is mainly created through natural forms. Plantings and various landscape feature spaces express a delicate and natural atmosphere.

In the display area, the designers respond to the buildings' Neo-Asian style and create a unique landscape feeling through hierarchy of planting, pierced feature wall, sunshine lawn, reflection pool and gravel landscape, etc. Adjacent to the waterfront landscape, the project provides residents a resort-style living space with combination of waterfront lifestyle and new interpretation of Chinese traditional garden.

Main Nodes Design

Entrance Plaza and Front Plaza of Clubhouse: The clean and quiet entrance plaza leads people to the clubhouse with a bluestone paved path flanked by rows of bamboos. A reflection pool responds to the buildings' facades and the aquatic plants pat the water surface with breeze. Gravel strips spread in reflection pool, paths and lawn. Light and shadow play an interesting effect through the pierced feature wall to the well-paved ground.

Sunshine Lawn and Parking Lot: The whole parking lot is enclosed by lush greenery in various heights. Hidden in plants, the parking lot enjoys an ecological and natural environment. The sunshine lawn surrounded by trees and shrubs brings a hint of leisure for the whole space. Gravel paths meander in the bamboo grove and are spotted through sunshine and shadows. Walking in the grove aimlessly, one could enjoy a relaxed feeling.

Waterfront Landscape: The riverway separates the sales centre and the project. A wooden deck links the two sides, which is accented by weeping willows along the river. Seen from the clubhouse to the opposite side, the buildings in layered

1

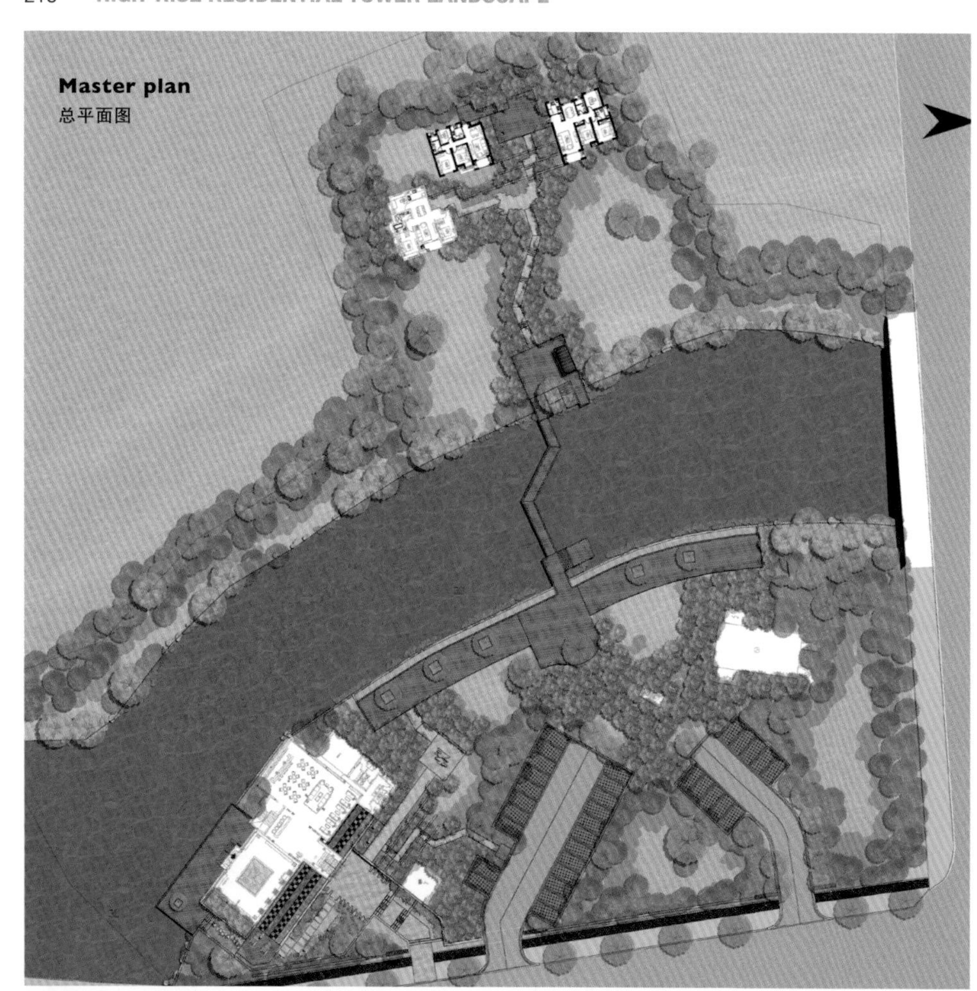
Master plan
总平面图

planting express a unique spatial feeling. With clusters of aquatic plants spreading along the river, the waterfront provides a fresh and natural atmosphere.

Model Housing Display Area: Walking through the wooden deck, one will enjoy a panoramic view of the lake. In this display area, one is fully immersed in a beautiful mood. Lake View Settlements provide residents an elegant lifestyle with lush greenery and natural feeling.

1. A delicate and natural atmosphere
2. The clean and quiet entrance plaza
3. A bluestone paved path flanked by rows of bamboos
4. Layered planting

1. 精致自然的空间
2. 简约静谧的入口广场
3. 青石道路两侧伴着整齐树立的碧竹
4. 植物层级丰富

该项目在整个园区中，设计师从功能上将南北区进行划分，使其形成不同的空间感和体验感。北区集中体现功能性空间，形成功能完善、大气尊贵的空间效果。南区主要以自然空间形态来营造，通过自然式的植物种植，多种小空间营造等，形成南区精致自然的空间感受。

在展示区部分，设计师以建筑的新亚洲风格为基础，用层级的植物空间、镂空景墙、阳光草坪、镜面水景、砾石景观等诠释这一风格下的独特景观感受。整个项目紧邻滨水景观带，如此看来，这正是一座富有水岸生活与现代新古典园林特色的滨水度假型生活府邸。

主要节点设计

入口广场与会所前广场：简约静谧的入口广场将人群引入会所区域，青石道路两侧伴着整齐树立的碧竹。一轮镜面水浅浅的与建筑交相呼应，水面上几丛水生植物在微风的作用下轻抚水面。条状砾石蔓延在镜面水中、青石路面上、草坪间，透过镂空的建筑影壁，光和影正作用在被铺装划分的刚刚好的地面上。

阳光草坪与停车场：整个停车场区域被高矮不同的层层绿植所包裹，完全隐蔽在植物中的停车场生态自然、植物层级丰富，乔灌木环绕的阳光草坪为整个空间透出一丝轻松感受。碎石汀步在竹林中回转，阳光透过竹林顶部洒下，斑驳的光影投射在迂回的小路上。即使在其中闲走，也会有种休闲自得的感受。

3

4

5

6

7

8

9

滨河景观带：河道将售楼处与项目区分隔开，一条木栈道连接并贯穿两岸，沿河两岸垂柳的枝叶垂抚水面。从会所区向河对岸望去，建筑在层叠的植被空间中衬托出一种别样的空间感受。丛生的水生植物在河岸两侧堆积蔓延，沿河岸走走，微风拂面、碧水清清，新鲜的绿意就这样在感受中催生。

样板房区域：通过长长的木栈道，进入到样板房区域，沿木栈道一路走来，湖边美景尽收眼底。仅仅伫立在河对岸片刻，已经完全沉浸在整个区域的美好感受中。居住在这里，生活与绿意融为一体，雅致生活就这样应运而生。

5. The path with green plants and flowers
6. The architecture on the opposite site of the river
7. The stairs on the slope
8. The water feature in the model housing display area
9. Night view of the water feature

5. 小路两旁种有绿色植被和花朵
6. 河对岸的建筑
7. 坡路台阶
8. 板房区域水景
9. 水景夜景

ZHONGZHOU CENTRAL PARK, SHENZHEN

深圳中洲中央公园

Location: Shenzhen, China
Completion: 2014
Design: SED Landscape Architects Co.,LTD.
Photography: SED Landscape Architects Co.,LTD.
Area: 90,800 sqm

项目地点：中国，深圳
完成时间：2014年
设计师：SED新西林景观国际
摄影：SED新西林景观国际
面积：90,800平方米

Zhongzhou Central Park is one of the "Central Parks" produced by Zhongzhou Estate and located in 26th zone of Bao'an district with an area of more than 90,000 square metres covering A, B and C districts. District A and B are high-grade residence while C is mainly for the comprehension of commercial and offices which takes up 510,000 square metres for the buildings. Being surrounded by the traffic nets and across from the park, it has a great geographical advantage.

Confronting both the urban development and the vintage monument, the project chose Art Deco to express this sort of cultural shock to reveal the historical sediment and natural treasure. Art Deco originates from France between the artists and the nobles who urged to show their life style and avant-garde attitude, which is kind of similar with those nowadays. That is also the most charming part of the Art Deco. The project pays attention on the texture and gloss of the materials with the stereotype from the geometric art symbols in Art Deco to create special aesthetics ambience. Featured sculptures of elegance together with the brief space arrangement offer a character landscape design.

Adorned with the Nobleness and Graceful Quality

District A, being abounded by Art Deco feelings, lies in the northwest part with an area of 40,000 square metres in total. Drained from the natural source, the inner water system is separates into two parts respectively as the pool and lake for the golf island by the axis. The landscape spots are joined by elegant line smoothly. Open leisure space, semi-open green area and private garden form an eco-system in the district. Innovation ideas has been used to combine the SPA resort, vertical diversity and the landscape together without any deliberation, expressing the grace at the same time.

Diamond Dynasty, the grand entrance, manages the front garden and meanwhile is the important mark of the project. Its modern fashioned details, gorgeously decorated, show a ceremony feeling in royalty naturally. Ordered design makes people feel slowing down by the gentle elegance.

1-2. Bird eye view of the community
3. The entrance
4. Night view of the community landscape

1、2. 住区景观鸟瞰图
3. 入口
4. 小区景观夜景

1

2

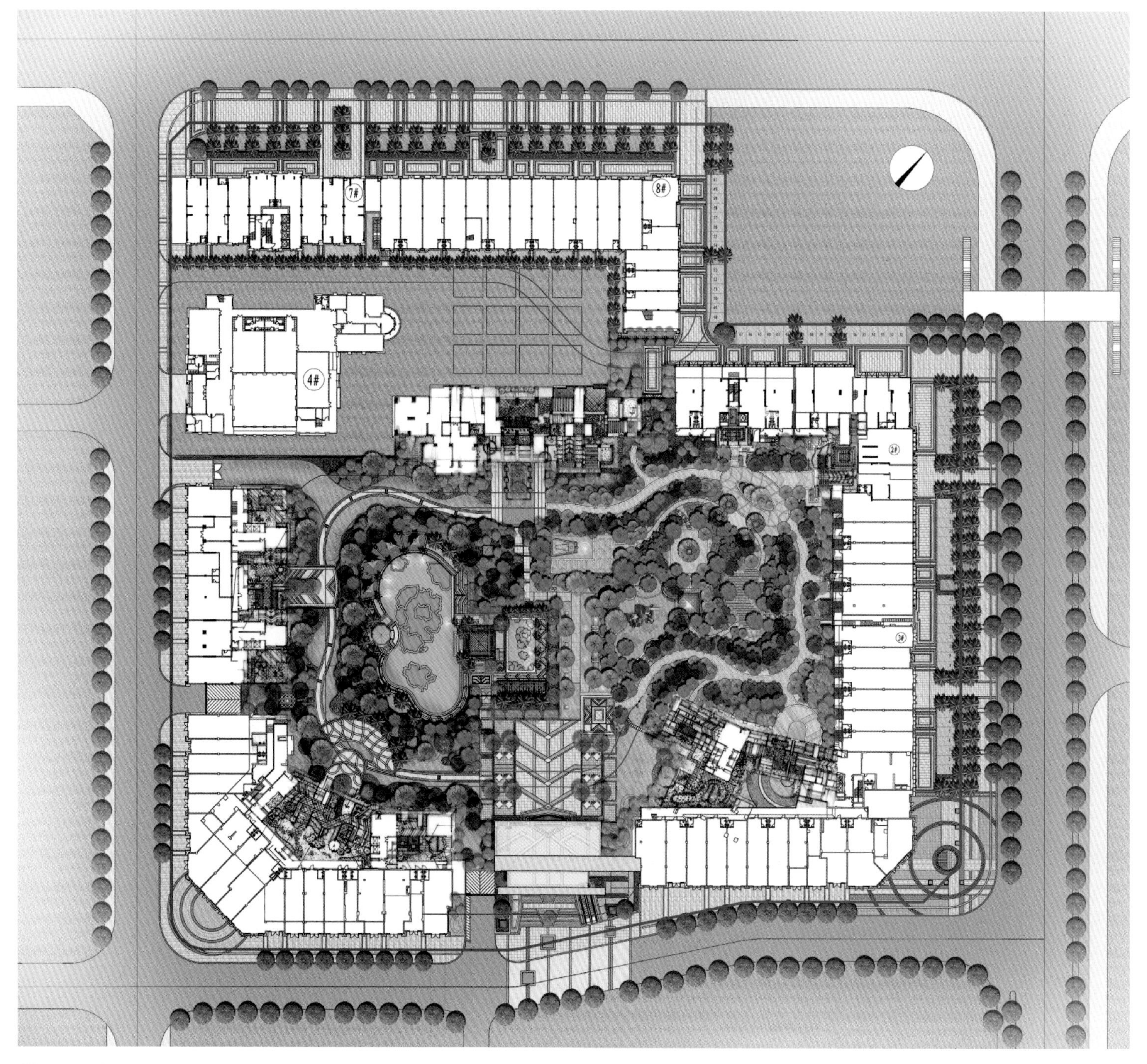

Master plan
总平面图

Cuishan court section
翠山雅苑剖面图

Passing through the Diamond Dynasty, a plaza shows up named after the sun that is the typical sign of reign's power and sparkling life. The real luxury from the nobleness brings the residents a high end identity and the construction of the pretty platform and the party square is to highlight and fit it.

The "Time Rose Pool", moreover, lets people fully relaxed in the peaceful mood. The rose pattern dose not mean the beauty of this flower only but the joy, the entertainment and the blooming life representing the memory of time and lasts forever, becoming the "rose of time".

Following the function parts the pool is divided into the children's part and the adults' part. The whole pool is managed in close-end way and raised up 750mm. The fitting room takes use of the underground space accompanied with a hotel scale of the SPA resort and the bathrooms.

The Wizard of Oz is a roof garden within which the plants rise layer by layer, alleviating the depression feeling brought by the high-rise buildings and creating a sense of romantic with interest. The architecture fits the city figure and shows the expanding and rich room, the surrounding source is under proper use with protection, everyone can feel the friendly atmosphere and attached to their homeland...these are what we all expect.

3

4

Geomantic Vegetation Arrangement

The vegetation design not only makes the living circumstance comfortable and delicate but also takes care of the variety of the space to make it feel more natural flowing. The shape-followed plants layout gives a rhythm. According to the geomantic sayings ever green trees are priority in the main entrance with the front scene consisted of broadleaf plants covering the whole Diamond Dynasty, explaining the traditional auspicious implies. Linkage and screen are set to complete the whole geomantic pattern as well.

Exalted Residential Impression Made by Sculptures

Eagle-shaped sculpture is taken in use as a mark of alien aroma. Art Deco was populated in USA especially its heart city, New York. As the result their bird of union, eagle, has a unique representation. Metalised bronze surface and its abstract looking with mechanical aesthetics show a kind of distance exalted feeling.

Art Details in Pavements

Pavements are almost in warm colour with Art Deco style showing the texture and the sheen. Intense contrast, geometric figures and pure colour mixed by metal feelings give a splendid visual impact.

5

宝城26区中洲中央公园是中洲地产精心打造的“中央公园”系列的经典之一。该项目位于宝安区宝城26区，一期项目总占地面积为9万多平方米，分A、B、C三个区，其中A、B区为高档尊品住宅，C区为集中商业及办公的城市综合体，总建筑面积51万多平方米。西临前进路，东沿公园路，北为裕安东路，南接创业东路，5号地铁和规划中的10号地铁在此交汇，两处地铁出入口将与项目的地下商城连通，南侧与灵芝公园隔路相望，环境优美，地理位置十分优越。

宝安发源地在都市重建计划下，一面是城市复兴，一面是怀旧情怀，唯有装饰艺术风格能完美表达这样的文化冲撞——百年艺韵积淀与绝版人文地脉相契，潮流来去，唯经典永恒。装饰艺术起源于法国，是从新古典过渡到现代主义之间的一种装饰艺术风格，它的风行伴随着新一代富裕阶层的诞生，这些新人群渴望表达自己的生活态度，确立自己的生活方式，所以当他们既不想回归传统，也不愿意陷入冰冷的机器或网络世界里时，这时装饰艺术就成了一种极佳的语言。装饰艺术的无限魅力，就在于对装饰淋漓尽致的运用，且不论时代如何变迁，都能在其中出现新突破。

装饰艺术贵族气派生活，典雅质量

以装饰艺术、新装饰主义风情为蓝本精心打造的中央公园A区景观，位于项目的西北端，由砖石王朝、太阳广场、时间玫瑰、水岛高尔夫、迭翠台等多个特色的景观节点组成。整体面积约4万平方米，A区源引自然的内部景观水系、以中轴为分界线，主要分为泳池景观区和水岛高尔夫两个板块，形成七个功能性景观节点(一条主轴、两个景观中心，七个功能性景观节点)，用优雅而流畅的线条串联各景观节点，开放的公共休闲空间、半开放绿地组团园林、高私密度的私家庭院，形成一个园区内的绿色生态系统。采用丰富的设计元素，将SPA度假享受及立体空间无缝景观连接概念同时引进，并且通过精巧的设计和安排，形成别具一格的景观环境，也营造出一种尊崇的贵族气派。

砖石王朝是一个气势恢宏的典雅入口前庭，既起到前庭管理口的作用，同时也是楼盘的重要标识。现代时尚而注重细节，加以华丽而艺术的装饰，一种皇家般尊荣的仪式感开始上演，严谨有条理的秩序，使得都市快节奏的生活步调瞬间转换至一种优雅徐缓的高贵格调中，充满仪仗感的风景带给居者宛若君主的礼遇。

穿越砖石王朝，来到soleil广场，以地面上太阳而命名，太阳在古代西方是帝王令牌上的典型标志，象征着权力和闪耀的人生。真正的奢华，在于专属的尊贵感。为契合居住者的高端身份，营造出真正高规格的社交平台，会所外设置聚会广场，复活真正优雅的奢华生活。

在尊品豪宅体验中，还不得不提到归家式度假

6

7

享受“时间玫瑰泳池”。有一种努力，比任何快乐都更恒定，更温暖，更让人在安宁和祥和中享受心灵和思想的盛宴;有一种美和时间一样持久，一样深沉，更像玫瑰一样绽放。此时，泳池中的玫瑰不再是一种图案，是游嬉，是欢乐，是争艳绽放的生命，融入了时光的记忆，顺着时间往下延续，永恒。成为一种“时间的玫瑰”。

根据功能分区，分为100平方米的0.5米深儿童池，700平方米0.8～1.5米深的成人池，泳池为封闭式管理，整体抬高0.75米。最大的特点是，更衣室利用地下室空间，整体比泳池道路下挖1.5米，高2.7米，室内净空2.35米。上端景观廊架上抬1.2米，总体相对地库顶板上抬3.2米，设置有坐厕、换衣间、洗手盆及淋浴间，形式酒店式立体的度假SPA享受。

绿野仙踪/迭翠台是一片屋顶花园，绿色的植物层层茂密，次第上升，不但弱化了高层建筑

Leimeng lake garden sections
雷蒙湖水景园剖面图

5. The path with green plantings
6. The water feature
7. Night view of the water feture

5. 种有绿色植物的小路
6. 水景
7. 水景夜景

给居者的压抑感，还营造一种趣味性和浪漫优雅。我们憧憬这样的景象：这里未来的建筑，与城市贴切的融合，并体现出空间的延展与丰富；对周围环境、资源合理的享用，而又谦和的保护；每一个居住或路过的人，都感受到一种亲切、随和的场所氛围，油然生出依恋和归属……城市、土地、家园，在这里完美的相遇。

藏风显水，风水植物空间格局

植物景观设计将回归自然的现代生活理念和异国风情相结合，创造精致典雅，极具人文气息的人居环境。一方面注重空间的变化，营造收放有序的自然空间。另一方面，植物根据设计构图形式布置，突出流线的韵律，彰显现代住区风情的优雅。同时在植物的运用及选型中，注重风水的说法，在主入口尽量选用常绿、阔叶植物为前景，把整体的砖石王朝掩盖在大树中，体现了传统的长盛不衰，大隐的吉祥寓意。在空间的转换中，也注重强调衔接及障景

8. The fountain nozzle　8. 喷泉喷嘴
9. The sculpture　9. 雕像
10. The shelter　10. 凉亭
11. The playground　11. 游乐场

的应用，让整个空间形成收放有致，藏风显水的风水植物格局。

艺术雕塑和小品，置身尊贵府邸

艺术雕塑小品上，选用具有装饰韵味的风情元素为主符号，如鹰雕塑。装饰艺术，发源于法国，而兴盛于美国，世界的装饰艺术中心即在纽约，作为国鸟的鹰自然有其独特的代表性。带有金属味的古铜色肌理及抽象造型，具有机械美学感觉的几何线条，同时产生富有距离的尊贵感。

装饰性铺装，艺术细节

该项目艺术的铺装细节主要以暖色调为主,融入装饰艺术风格，注重表现材料的质感及光泽，以荔枝面、烧面及光面的面材处理方式形成质感对比，造型设计中多采用几何形状或用折线进行装饰；色彩设计中强调运用鲜艳的纯色、对比色和金属色，造成华美绚烂的视觉印象。

• *Low-rise villa landscape*

别墅居住区景观

CHELONA LANDSCAPE
舍罗那景观

Location: Prachuap khiri khan,Thainland
Completion: 2013
Design: Wannaporn Suwannatrai
(Openbox Company Limited)
Photography: Wison Tungthunya
Area: 2,740sqm

项目地点：泰国，班武里府
完成时间：2013年
设计师：瓦那邦·苏万纳特来（OPNBX设计公司）
摄影：维森·东新雅
面积：2,740平方米

OPNBX was approached by Sansiri to propose landscape design for a beachfront condominium in Huahin. Developed to be a weekend home, Chelona has been assigned to be resemblance of a relaxing, little town in Greece. In close collaboration with Agalio Studio, the design architect of the project, architecture and landscape concept has merged seamlessly into one.

The site is a typical beachfront formation with a narrow frontage facing the beach and very long strip stretching over ten times the width to connect to the access road. The site was mainly covered in sand. One third of the land has been occupied by fully grown existing trees. All of them are local and have withstood sea wind for a long time and have turned into sculpture forms.

Typically for a long site with a narrow side facing the beach, only a few units have the direct sea view. The rest of the units would have to rely on variety of intense landscape creation to become the views. Even with limited palette of white to light colour hardscape, water feature and natural plant colours, variation of experience could still be created, based on differences of directions, lines, curves and height.

Aside from the obvious thematic design, the most important task is to keep as many of the existing trees as possible. OPNBX team has begun by documenting all trees, going through selection process, and propose tree saving features in design: such as

- Adjusting the design to fit the existing tree locations, not the other way around. In a few extreme cases, positions of buildings are adjusted to fit in among the existing trees;
- Leaving lots of space around existing balls;
- Hiring a specialist to be on board since day one of construction;
- All these have to be done without losing the efficiency of the salable.

In this case, the designers considered themselves very successful in saving the trees and they return the favour by making the project feel very "lush" and "alive" since the first day of opening.

1

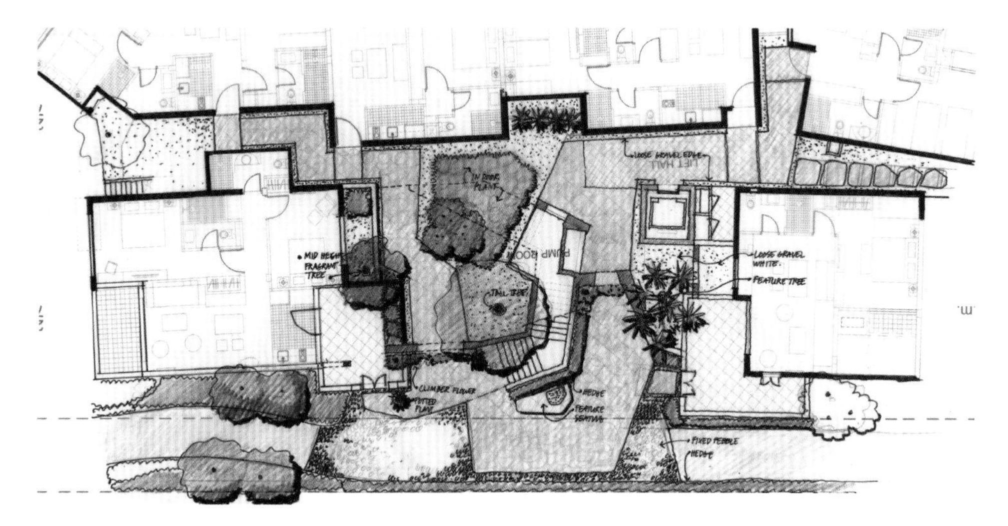

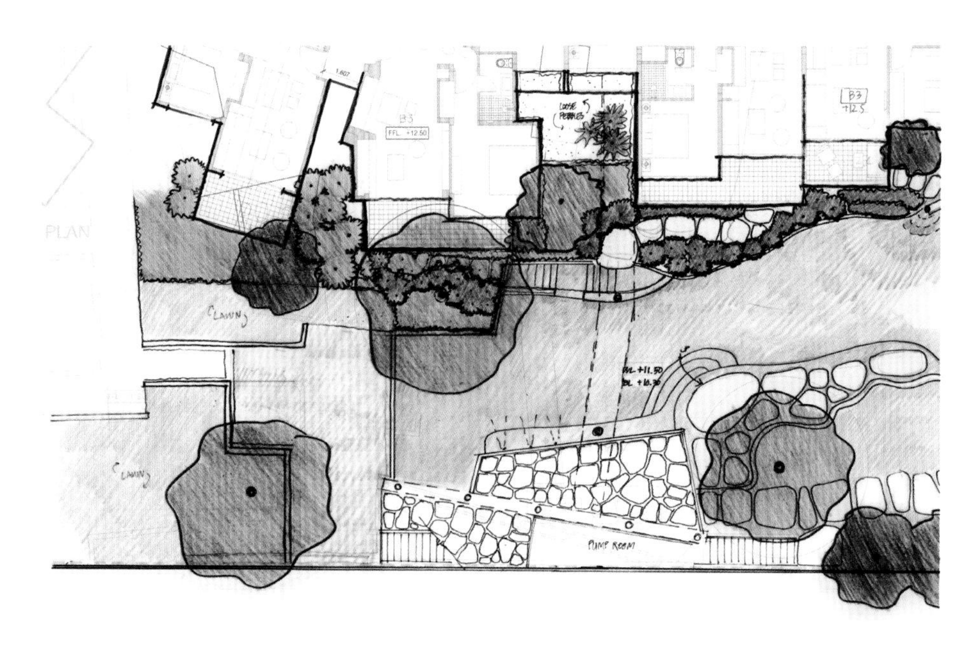

OPNBX设计公司受委托为泰国华欣的一处海滨公寓进行景观设计。舍罗那小区的定位是度假公寓，其设计类似于一处轻松的希腊小镇。景观设计师与建筑设计师阿加里欧工作室紧密合作，将建筑与景观概念合二为一。

项目场地为一个典型的海滨地形，狭窄的前庭正对海滩，由一条狭长的小路与入口通道相连。整个场地主要由沙子覆盖。三分之一的土地上已经栽种了成熟的树木，这些本土树木在长时间抵御海风的过程中形成了富有雕塑感的造型。

因为居住区只有一个窄边朝向海滩，仅有几户享有直接的海景。其余的住宅单元只能依靠丰富的景观特征来优化视野。尽管硬景观的色彩仅限于白色和一些浅色，水景和自然植物的色彩仍能以不同的方向、线条、曲线和高度为人们带来丰富的体验。

除了明显的主题设计，最重要的任务就是尽可能多的保留原有树木。OPNBX的设计团队为所有树木都进行了记录，通过挑选流程，在设计中考虑了树木的保护措施，例如：

·调整设计使其适应原有树木的位置，而不是调整树木适应设计。在一些极端的案例中，建筑的位置甚至参考树木的位置进行了调整

·在树木周围保留了大量空间

·聘请专家参与设计流程，直至开始施工

·这些措施的实施不能影响公寓的销售效率

在本案中，设计师出色地完成了树木保护任务，而树木则回报给项目郁郁葱葱的景色。

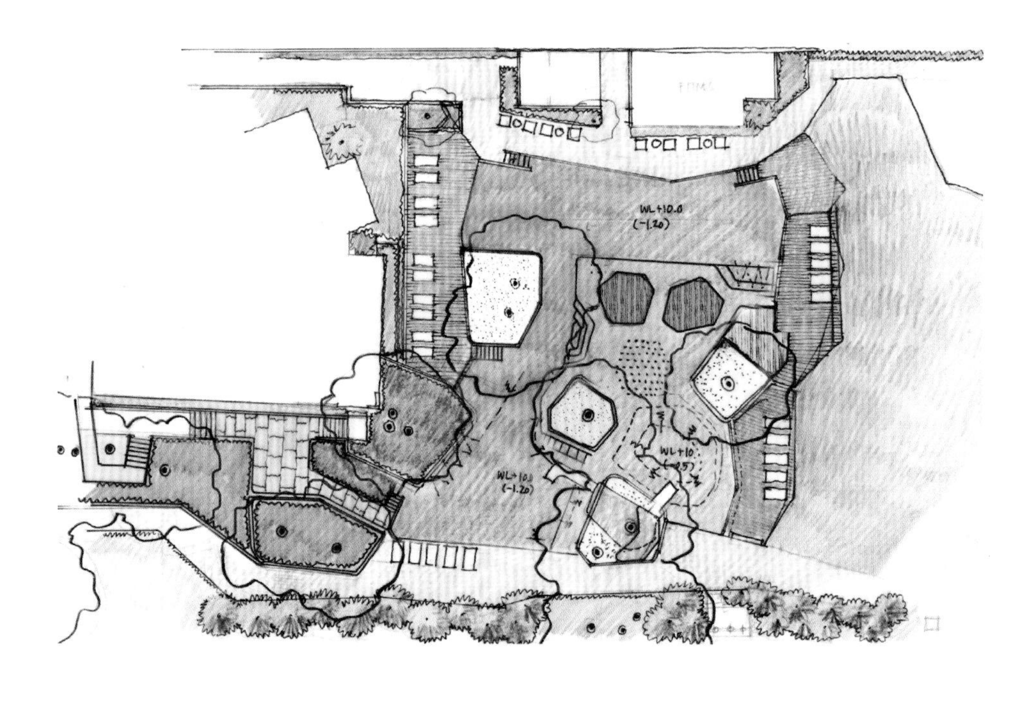

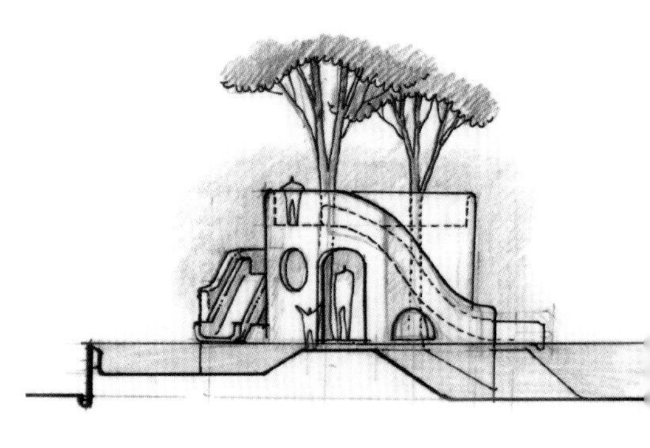

1.Architecture and landscape concept has merged seamlessly into one
2.A narrow frontage facing the beach
2.The slide for children

1. 建筑与景观合二为一
2. 狭窄的前庭正对海滩
3. 儿童们做游戏的滑梯

4

4. The water feature and the lush green plants
5. The trees are local and have withstood sea wind for a long time and have turned into sculpture forms
6. Variety of intense landscape creation
7. The clear water

4. 水景和茂密的绿色植物
5. 本土树木在长时间抵御海风的过程中形成了富有雕塑感的造型
6. 丰富的景观元素
7. 清澈的池水

INDOCHINA VILLAS SAIGON

Indochina西贡精品别墅

Location: Ho Chi Minh City, Vietnam
Completion: 2014
Design: ONE landscape
Photography: Jason Findley, Aaron Joel Santos
Area: 80,000sqm
Plant information: Plumeria species, Terminalia mantaly "Variegata", Delonix regia

项目地点：越南，胡志明市
完成时间：2014年
设计师：ONE景观设计公司
摄影：詹森・芬德利、亚伦・乔尔・桑托斯
面积：80,000平方米
植物信息：鸡蛋花、小叶榄仁"斑锦"、凤凰树

Developed by Vietnam's premier real estate company INDOCHINA LAND, boutique waterfront residential project in Ho Chi Minh City is an upcoming high-end villa development of 8 ha. Inspired by the linearity of the vast paddy fields that marks Vietnam's dramatic landscape, the design includes a series of vibrant community spaces that are interconnected through a safe and secure public realm based on "access for all" philosophy. The community spaces – a linear waterfront park, community swimming pool plaza, neighbourhood park and children's play park are all linked together with the unifying geometrical abstraction of rice grain, thus an extension of the paddy fields concept.

This villa experience begins with the entry statement at its frontage to the main road. Inspired by the rice pattern the boundary wall is unique in creating a strong identity and character. The boundary wall further merges with the art wall – the most important feature of the entry experience. The wall with its sculpted pattern with deep shadows changes character in the evening when the light box installed behind the wall softly glows to create a dramatic statement.

The rice inspired pattern unifies and manifests itself in a series of scales throughout the project especially in the community pool area located just after the entrance. Designed as the key gathering place, this area visually connects the adjacent lake with its infinity swimming pool featuring a linear deck canopy structure. With its rich patterned panels as both horizontal and vertical support elements it is an elegant structure of sculptural qualities. A shaded corridor of similar design character connects the Tai Chi Plaza with the waterfront BBQ deck and features children's water fountain. Inspired by the lotus and conceived as a feature play fountain the sculptural installation is more an artwork than a water feature. Dramatic lighting in this area highlights the sculptural elements and brings a boutique resort feel to the development.

Further down the development is the neighbourhood park, the other key park space within the project. Conceived as green oasis and also accessed from the

1. Night view of Eden
2. The entrance
3. The wall with its sculpted pattern with deep shadows changes character in the evening
4. Clean and simple landscape

1. 伊甸园别墅夜景
2. 入口
3. 夜晚，带有镂空图案的墙壁将通过后置灯箱发出柔和的光芒
4. 简洁干净的景观

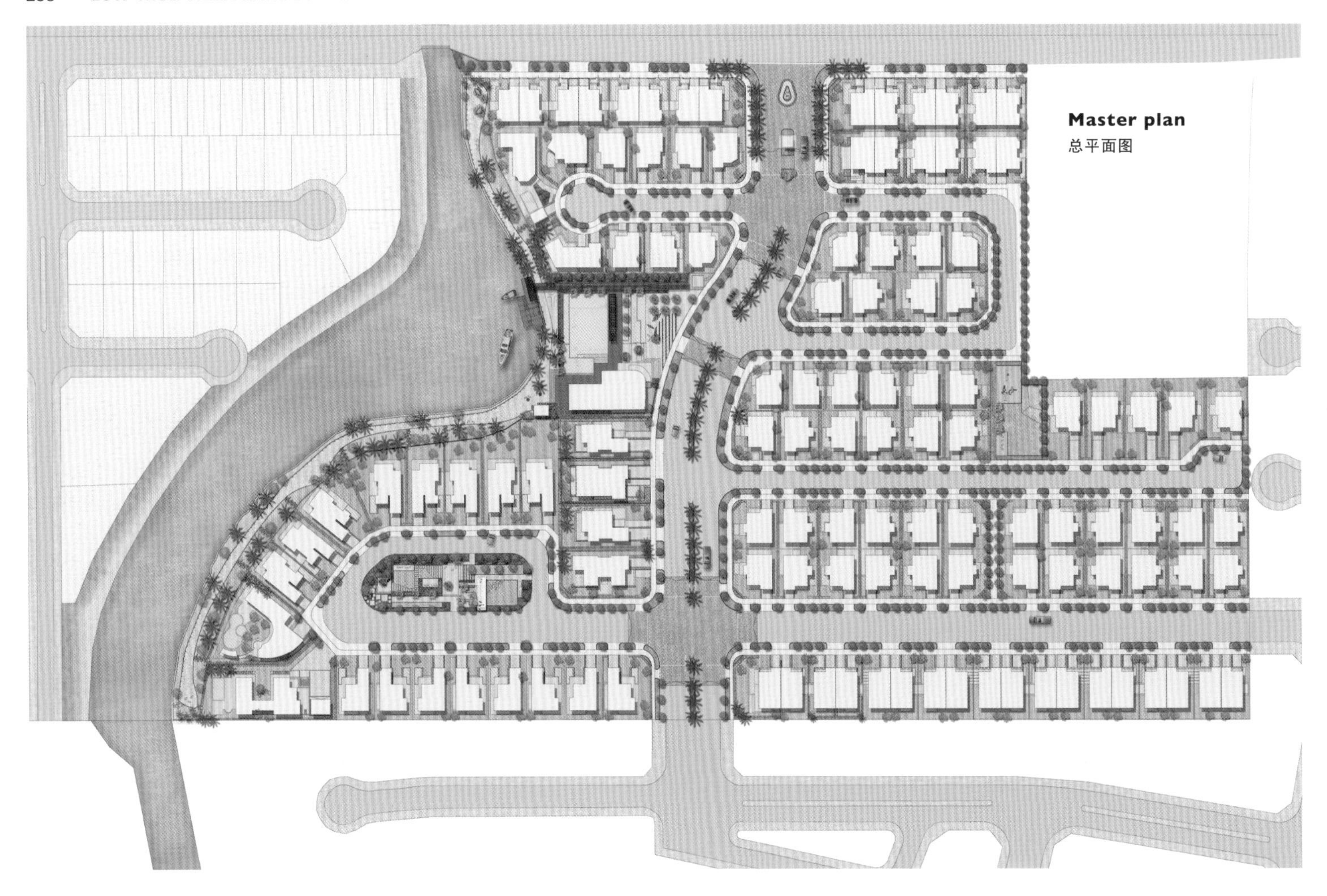

Master plan
总平面图

linear waterfront park, it is a space with many unique features. The shape and pattern inspired by the geometrical abstraction of the rice grain finds new forms, the most common being the feature screen. Set against lush planting the screens create a new setting within this park. A series of platforms with linear benches act as places for relaxation and contemplation. The park also contains a central floating canopy hovering among the tree layer, its elegant floating plane in complete contrast to the tree foliage. Supported by a series of slender metal columns its sculptural quality is further enhanced by a series of cutouts in shapes of the pattern, not only to reduce load but also invite exciting play of shadows on the ground. A series of stepping stones on the lawn in the same geometric shape adds unity to the theme and creates the illusion of the pavilion cutouts being are mere subtractions only to appear on the ground as stepping stones. A series of sculptural seats as extrusions of the geometric shape appear among the greenery as contrast against the backdrop of lush foliage. Lighting further creates drama in highlighting these key landscape features within the park.

ONE's approach of integrating art and culture in this project creates a unique landscape to be enjoyed both the residents and its visitors, making EDEN a landmark development of Ho Chi Minh City and creating a new benching in resort style city living.

3

4

由越南首屈一指的房地产公司印度支那地产开发的伊甸园精品别墅项目位于越南胡志明市，基地定位为高端别墅区，占地8公顷。设计灵感来自广阔的稻田，标志着越南传统特色的田野景观的线性。设计将一系列充满活力的社区空间，通过一条安全可靠又方便易达的公共走廊联系起来。社区空间——线型的海滨公园、广场、社区游泳池、社区公园及儿童游乐场地等元素联系整合，架构出统一的社区公共空间系统，成功地演绎了越南传统的水稻梯田景观。

该项目的体验从与主路相通的正门开始。边界墙从米粒图案中获得了启发，呈现出强烈的形象和个性。边界墙进一步与艺术墙——入口体验中最重要的元素融合起来。夜晚，带有镂空图案的墙壁将通过后置灯箱发出柔和的光芒，形成戏剧化的效果。

米粒图案在项目中反复出现，特别是在紧邻入口的公共水池区尤为明显。作为主要的集会场所，这一区域在视觉上通过无边界游泳池与旁边的湖泊联系起来。游泳池以遮阳露台结构为特色，带有米粒图案的支承元素构建了优雅的

5

6

5.The pavilion
6.Set against lush planting the screens create a new setting within this park
7.The infinity swimming pool featuring a linear deck canopy structure
8.The pavement
9.A series of platforms with linear benches act as places for relaxation and contemplation

5. 凉亭
6. 映衬在繁茂植物直线的屏风为公园增添了新设施
7. 游泳池以遮阳露台结构为特色
8. 铺装
9. 一系列带有长椅的平台是人们休憩沉思的场所

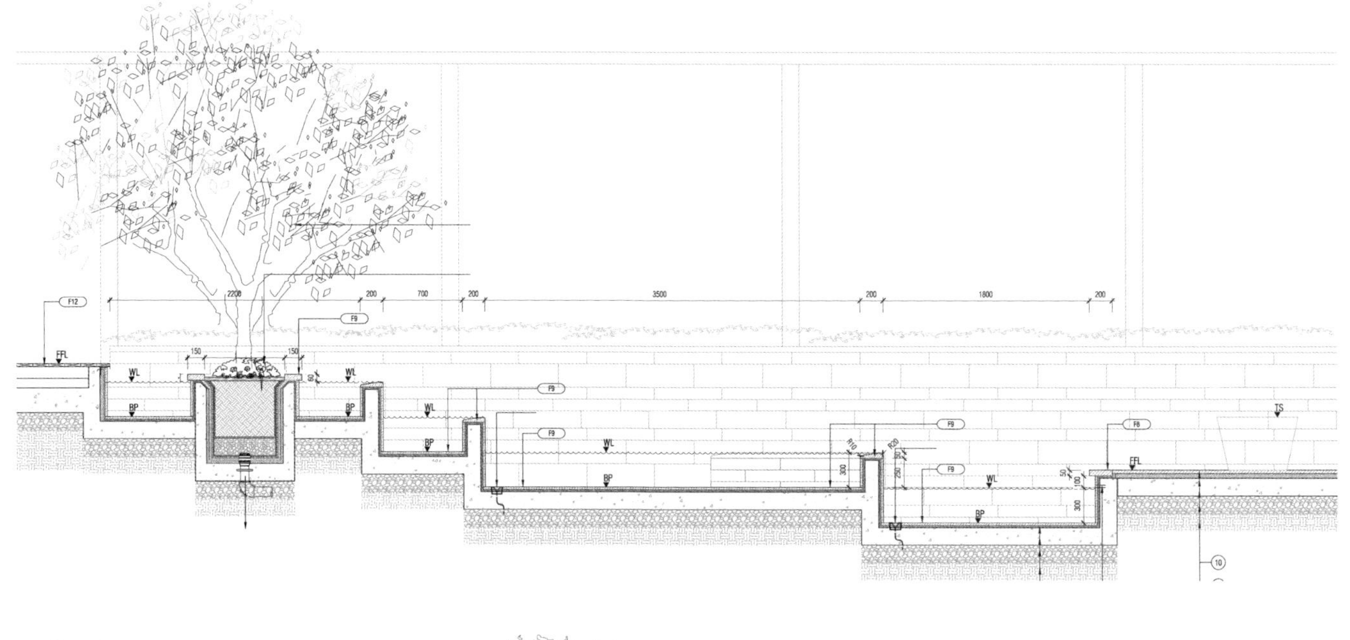

Sections
剖面图

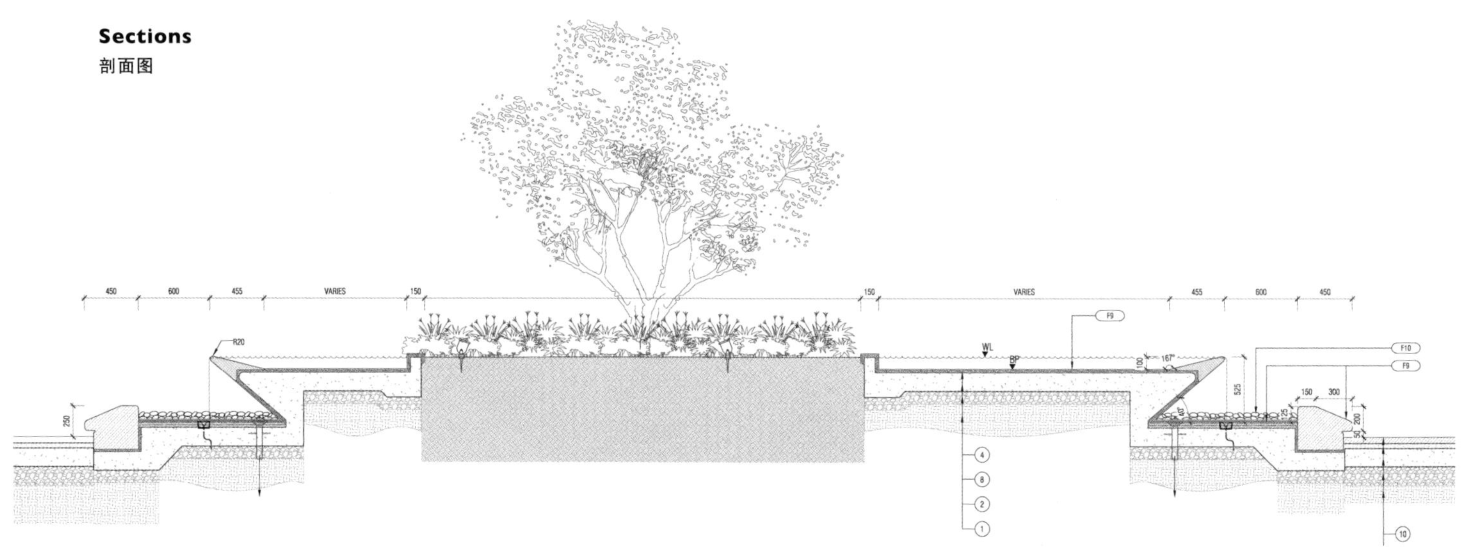

雕塑感。采用类似设计的长廊将太极广场与滨水烧烤区连接起来，以儿童喷水池为特色。特色喷泉从莲蓬中获得了灵感，更像是一件艺术品。这一区域的动感灯光突出了雕刻元素，为整个住宅区带来了精品度假村的感觉。

再向里走是社区公园——项目中另一个重要的公园空间。作为一片绿洲，它同样有滨水带状公园进入，拥有多个景观特征。米粒图案以新的形势呈现出来，最常见的就是特色屏风。映衬在繁茂植物直线的屏风为公园增添了新设施。一系列带有长椅的平台是人们休憩沉思的场所。公园还包含一个中央悬浮华盖，悬浮在树木之间。它优雅的平顶与树叶相互映衬。华盖由一系列细金属柱支撑，米粒剪影图案不仅减少了负重，还在地面上投下了有趣的光影。草坪上的踏脚石同样采用了米粒图案，实现了主题的统一，让人有一种凉亭剪影映射在地面上的错觉。呈现为米粒造型的雕塑座椅在绿植中与郁郁葱葱的叶子形成了对比。灯光则进一步突出了公园里的主要景观特征。

ONE景观设计公司的设计方案融合了艺术与文化，形成了独特的怡人景观，让伊甸园成为了胡志明市房产开发的地标，奠定了度假村式城市居住区的新标准。

10. Water feature
11. Water cascade
12. Central floating canopy
13. Inspired by the lotus and conceived as a feature play fountain the sculptural installation

10. 水景
11. 小瀑布
12. 中央悬浮华盖
13. 特色喷泉从莲蓬中获得了灵感

MARICOLLE

玛丽科勒日式花园

Location: Fukuoka, Japan
Design: wa-so design
Photography: Kazufumi Nakamura
Area: point A: 590.1sqm, point B: 197.3sqm
point C: 83.4sqm

项目地点：日本，福冈
设计师：wa-so设计事务所
摄影：中村和史
面积：A区590.1平方米，B区197.3平方米，
C区83.4平方米

This is a refurbishment project, with the garden and interior and exterior of wedding facilities located at seaside in Kokura city. It was requested to design not only beautiful but also as a place that could take a wedding party. We considered to create an extraordinary memorable and enjoyable space and scenery, therefore first of all, we assigned two concepts: "waterfront resort" and "highland resort". Existed grass area is expanded as a refurbishment area according to place together a part of parking, and a new garden taking a leading part of pool is made there. There are a bridge and an island covered with wooden deck as a stage for leading couple in ceremony. The sidewall of the pool is sloped to make the most of water lights efficiently and it makes sense for structure as well. Some water fountain nozzles are set up inline, hence they are possible to make a water wall. It is possible to control water height freely from a counter that was arranged inside area. There are the special spaces surrounded by plants on the other side, and wooden deck and tiled space are secured for the participants in ceremony as much as possible. On the other side, with the trees in wooden deck, it seems to be in a small forest when you participate in the party. We considered to use evergreen trees mainly for instance Fraxinus griffithii because the facility is open all the year round. Not only are there some exterior sofas, a projector and sound system are also furnished in this hall in addition.

该项目位于福冈市海滨地区，是针对庭院及婚礼服务场所室内外的翻修工程。设计师需要设计出不仅美观而且适合举办结婚典礼的花园景观。为了营造一个能给人带来难忘回忆，同时又舒适宜人的空间景观，设计师首先在设计中引入了两个概念："海滨度假胜地"与"高原度假胜地"。花园的池面上有一道小桥和一座小岛，小岛上覆盖着木制平台，婚礼中的夫妇会在这个平台上举行仪式。设计师通过水池的设计极大程度地增加了桥和岛与水景之间的联系，使得水池边缘的存在感尽可能降低。水池旁的边墙为倾斜式设计，在提高对水景灯光的利用率的同时，结构本身也别具特色。由于一些喷泉的喷口采用了内联的安置方式，形成了水墙景观。水墙的高度可以自由控制。在花园另一侧，是被植物环绕的特殊空间。在另一端，参加结婚典礼的宾客站在绿树环绕的木制平台上，仿若置身于一座微型森林。由于一年四季都有婚庆典礼举行，所以设计师主要选用了常绿植物比如光蜡树。花园中不仅安置了室外沙发，还设有投影仪和音响系统供典礼使用。

1

2

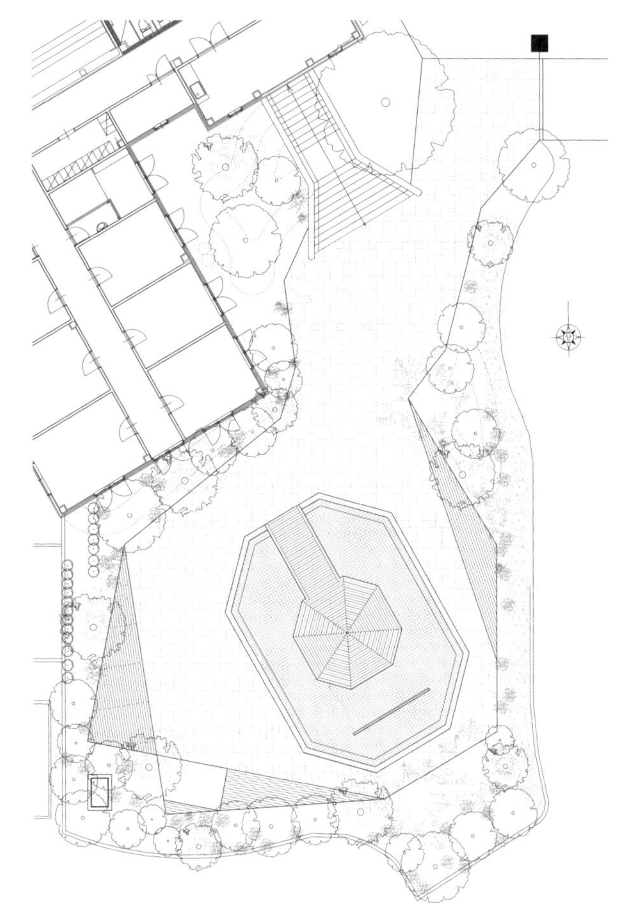

1-2. The featured swimming pool
3. The entrance area is mainly white and green
4. Exterior sofa for relaxing

1、2. 独具特色的游泳池
3. 入口以白绿为主色调
4. 可以休憩于此的室外沙发

4

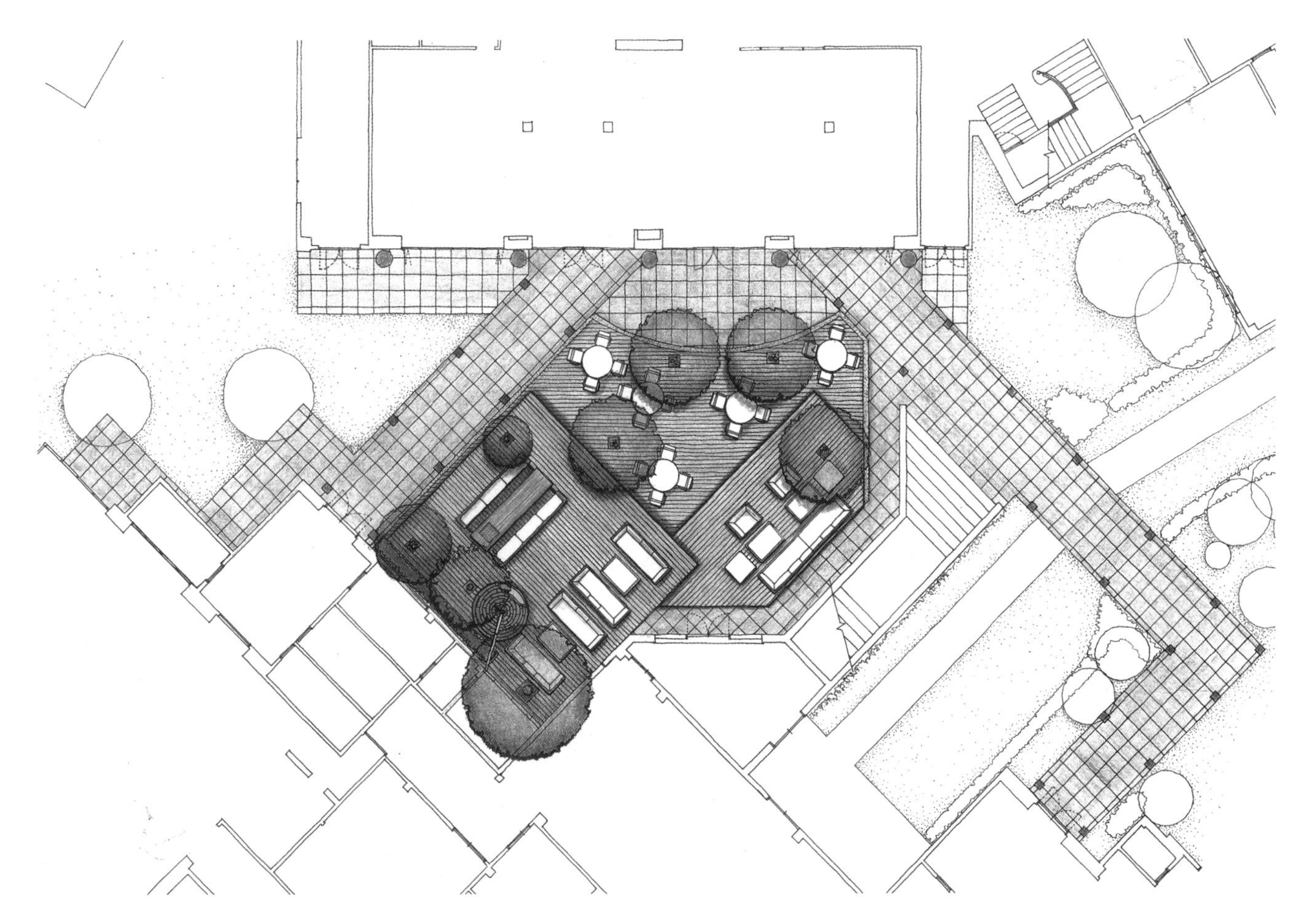

5. Night view of the landscape
6-7. lighting detail
8. Evergreen trees mainly for instance Fraxinus griffithii are used most
9. The pavement detail

5. 小区景观的夜景
6、7. 照明细部图
8. 设计中大量使用了常绿植物比如光蜡树
9. 铺装细部图

8

9

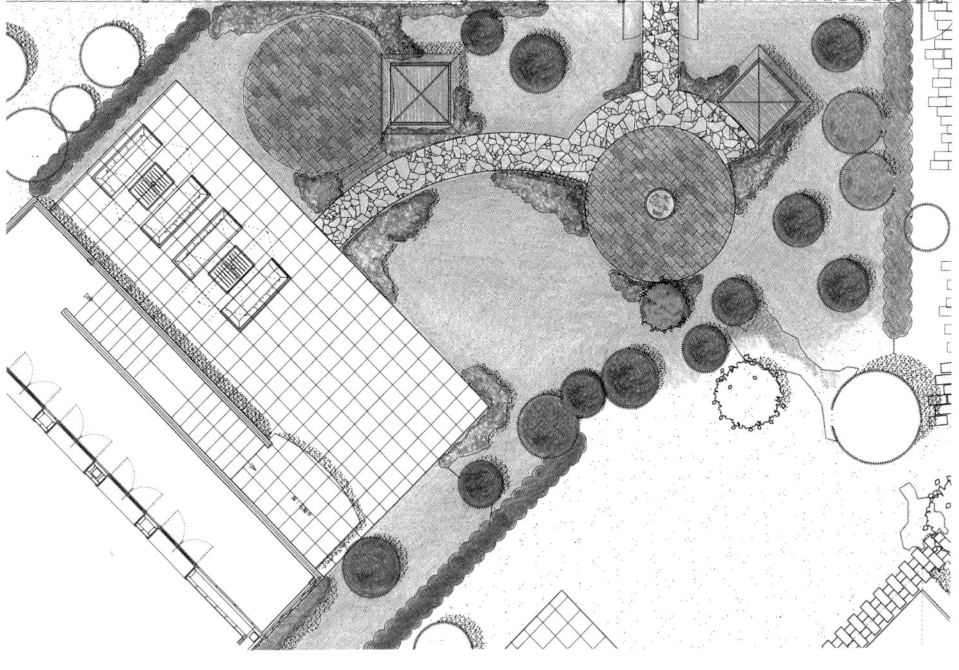

MANGROVE PARK AND NEWPORT QUAYS STAGE 1

红树林公园与新港码头项目一期工程

Location: Ethelton, Australia
Design: Taylor Cullity Lethlean
Photography: Ben Wrigley
Area: 7,000sqm

项目地点：澳大利亚，伊瑟尔顿
设计师：泰勒・库里提・莱斯林景观事务所
摄影：本・瑞格利
面积：7,000平方米

The redevelopment of the Port Adelaide inner harbour is one of South Australia's largest and most significant urban development projects. Taylor Cullity Lethlean was engaged by the Brookfield Multiplex – Urban Construct joint venture to undertake urban and landscape design for the first stage including the regeneration of the adjacent portions of Mangrove Park. Through close collaboration with Cox Architects, significant changes were made to the road treatments removing kerbs and creating a more pedestrian friendly environment.

Working within the constraints of the predetermined built form planning, Taylor Cullity Lethlean created a series of landscaped spaces of great variety and high amenity. The scope of services included the design of all exterior spaces including the waterfront promenade, outdoor structures and furniture, decking, paving and planting. Working within a tight budgetary framework, care was taken to add detail to furniture and structures wherever possible. Similarly, planting was selected to add patterning, colour and spatial definition to the landscape.

Soil conditions on the site are particularly difficult with the entire development constructed over remediated soils. Taylor Cullity Lethlean worked closely with geotechnical and civil engineers, soil scientists and arborists to develop a soil profile which met the stringent remediation requirements while providing a suitable growing medium for plants. The thriving plantings which have become a feature of the development are testament to the successful resolution of the extremely difficult site soil challenges.

In Mangrove Park at the southern end of the development, extensive soil amelioration and the planting of 37,000 indigenous plants has successfully rehabilitated this formerly degraded section of the park. Consultation with the adjacent school ensures that the park is a venue for their environmental education programs.

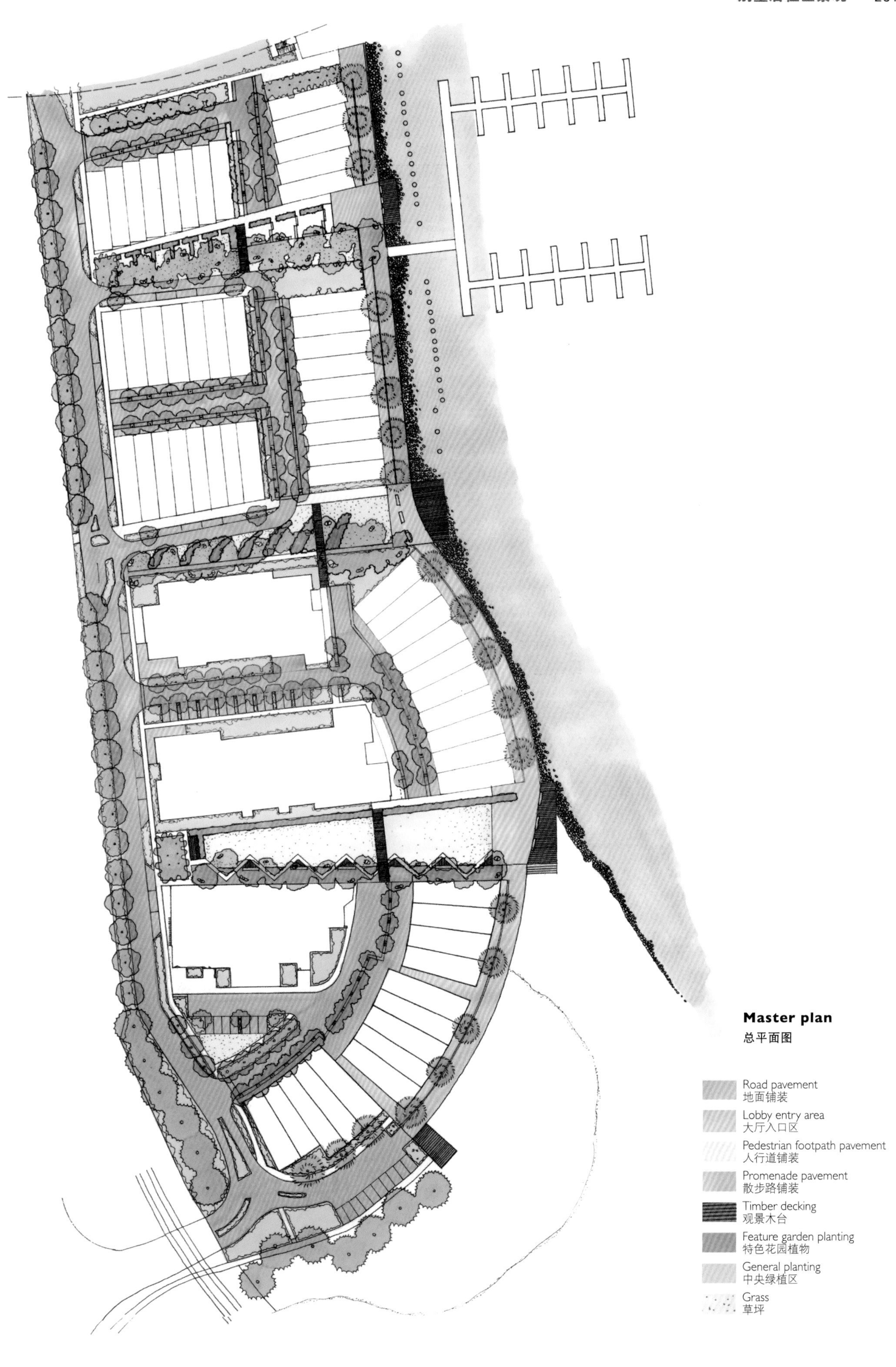

Master plan
总平面图

The first stage of Newport Quays provides residents and the general public with a readable hierarchy of private, communal and public spaces which contribute to the landscape amenity of the local neighbourhood and broader region. The rehabilitation of Mangrove Park and the design and construction of the first stage of the waterfront promenade with its associated decks, shelters and gathering spaces are particularly important aspects of the project for the enjoyment of the wider community.

Design of the residential landscape of the Newport Quays development involved an intensive collaboration with site architects and civil engineers to substantially influence the road character between residential buildings, including deletion of kerbs and adoption of centre drainage. Also, pavement selection and planting design were carefully undertaken with detailed review and consultation with all relevant parties. Particularly successful has been the planting selection to create a thriving landscape in dry coastal conditions in very limited soil volume.

The Newport Quays development is one of few examples in South Australia of intensively landscaped medium density housing and apartment developments in which a landscape architectural lead approach was taken to all outdoor spaces. The result has been a clear concept which expresses a hierarchy of public, semi-public and private spaces with public waterfront promenade and roadways, public/communal parkland wedges between building blocks, and private courts for townhouses.

The Newport Quays Stage 1 and Mangrove Park designs involved two major site specific environmental undertakings. The whole Stage 1 development was undertaken on a previously industrial brown-field site with considerable soil contamination. Extensive and thorough soil remediation was undertaken to allow residential development to occur. This involved intensive collaboration with engineers and scientists by the landscape architect to create a suitable growing environment for a successful landscape.

The second major undertaking was the rehabilitation and revegetation of a large portion of the Mangrove Park site. This site included a remnant stand of indigenous mangrove forest and associated samphire plant communities. The planting of over 37,000 indigenous plants has vastly improved and ensured the survival of the mangroves and samphire marshes and general biodiversity of this site on the Port River.

1. A series of landscaped spaces of great variety and high amenity
2. The thriving plantings which have become a feature of the development
3. The ornamental trees in front of the building

1. 灵活多变且高度舒适的景观空间
2. 欣欣向荣的植被绿化已经成了项目的标志性特征
3. 楼房前面种植的具有装饰性的小树

阿德莱德港的再开发工程师南澳大利亚最大、最富影响力的城市开发项目。泰勒・库里提・莱斯林景观事务所受布里克菲尔德市政合资公司委托为一期工程进行城市和景观设计，其中还包括对红树林公园的部分重建。通过与考克斯建筑事务所的紧密合作，道路的路边石被拆除，形成了更有利于步行的环境。

在既成的建筑形式规划的限制下，泰勒・库里提・莱斯林景观事务所打造了一系列灵活多变且高度舒适的景观空间，包括海滨散步道、户外街景设施、木板平台、地面铺装和植被绿化。由于项目预算比较紧张，街道设施和结构的细部设计经过了仔细挑选。同时，精选的植被绿色为景观增添了层次、色彩以及空间定义。

场地的土壤条件较差，整个开发项目都建在修复的土壤上。泰勒・库里提・莱斯林景观事务

4

所与地质技术工程师、市政工程师、土壤专家和树木栽培家通力合作，开发出了一种既能满足严格的修复要求，又能提供适合植物生长介质的土壤。欣欣向荣的植被绿化已经成了项目的标志性特征，证明了土壤改造的成功。

在项目南端的红树林公园，大面积土壤改良和37,000棵本土植物的种植成功地修复了公园的衰落部分。与周边学校的协商合作让公园变成了他们的环境教学项目的一部分。

新港码头项目一期工程为居民和公众提供了层次分明的私人、社区以及公共空间，有利于提升本地周边社区及相邻地区的景观舒适度。红树林公园的修复以及一期海滨散步道的设计（包括相关的亭台楼阁、集会场所）为更广泛的利益群体提供了更多的乐趣。

5

新港码头项目的居住区景观设计涉及了与建筑师与市政工程师的广泛合作，改变了道路与居民楼之间的关系，其中包括拆除路边石和采用中央排水系统等。此外，地面铺装的选择与绿化设计都经过了相关各方的详细探讨和磋商。项目最成功的地方就在于植被的选择，设计师在沿海干燥的条件下利用有限的土壤打造了欣欣向荣的景观。

新港码头项目是南澳大利亚少有的密集景观中密度住宅开发项目，其景观设计涉及了所有户外空间。项目的公共、半公共和私人空间层次分明，海滨散步道、道路、公共林地与住宅楼和联排别墅的私人庭院相互交错，融为一体。

新港码头项目一期工程与红树林公园的设计包含两块主要场地的特殊环境任务。整个一期开发都建在一块土壤受到大面积污染的工业棕地上。景观设计师与工程师和科学家共同合作，对场地进行了全面而广泛的土壤修复，保证了工程的顺利进行并提供了适合植物生长的景观环境。

第二个主要环境任务是对红树林公园的部分场地进行修复和植被恢复。该场地包含本土红树林和圣彼得草植被群。37,000余棵本土植物的种植提升并保证了红树林和圣彼得草沼泽的生存空间，同时也实现了港口河的生物多样化。

4. Particularly successful has been the planting selection to create a thriving landscape in dry coastal conditions in very limited soil volume
5. Night view of the community landscape
6. Planting was selected to add patterning, colour and spatial definition to the landscape
7. Care was taken to add detail to furniture and structures wherever possible

4. 设计师在沿海干燥的条件下利用有限的土壤打造了欣欣向荣的景观
5. 小区景观夜景
6. 精选的植被绿色为景观增添了层次、色彩以及空间定义
7. 街道设施和结构的细部设计经过了仔细挑选

SOMMETS-SUR-LE-FLEUVE

河上花园

Location: Montréal, Canada
Design: Williams Asselin Ackaoui & Associates Inc
Photography: Vincent Asselin
Area: 75,000sqm

项目地点：加拿大，蒙特利尔
设计师：威廉姆斯・阿塞林・阿卡奥伊景观事务所
摄影：文森特・阿塞林
面积：75,000平方米

The "Sommets-sur-le-Fleuve" project is a landscape design for a newly built prestigious apartment tower complex. Located on the shores of Nuns' Island, a small "green" island across the river from the Island of Montreal, the apartment towers are blessed with a sublime view over the St. Lawrence River, and the city of Montreal and its downtown.

A succession of four typologies of gardens takes us on a trip through pedestrian and bicycle paths, a pagoda band shelter, a sculpture garden and a formal square.

Lush planting, floral displays, butterfly gardens, tree-lined thoroughfare and formal gardens garnish and add value to the already beautiful site; they also embellish the residential complex, as well as offer a more agreeable view from the floors up, to the gardens at grade.

Plant Information

Trees: Amelanchier canadensis, Acer ginnala, Gleditsia triacanthos "Shademaster", Prunus serotina, Syringa reticulata "Ivory Silk", Thuja occidentalis "Fastigiata"

Shrubs: Actinidia kolomikta, Amelanchier canadensis "Balerina", Aronia melanocarpa "Vinking", Diervilla lonicera, Eonymus alatus "Compactus", Humulus lupulus, Hydrangea paniculata "Unique"

Perennials: Alchemilla mollis, Ajuga reptans 'Catlin's Giant', Aruncus dioicus 'Sylvestris', Calamagrostis x 'Karl Foster', Calamagrostis brachytricha, Cerastium tomentosum, Geranium sanguineum, Hemerocallis 'Joan Senior'

Master plan 总平面图

1. The swimming pool in the community garden
2. The green lawn
3. The lush planting

1. 小区花园的游泳池景色
2. 大片的绿色草地
3. 茂密的植被

"河上花园"项目是为一个新建的高档公寓小区所提供的景观设计。公寓楼坐落在与蒙特利尔岛隔河相望的努恩岛河畔，纵享圣劳伦斯河、蒙特利尔市及其市中心的美景。

四种类型的景观花园将引领人们穿过步行及骑行路径、弧形凉亭、雕塑花园以及广场。

茂密的植被、花卉展示、蝴蝶花园、林荫大道以及规划式庭园为已经十分美丽的场地锦上添花。它们还装饰了住宅小区，为楼上的住户和楼下的私人花园带来更宜人的景色。

植物信息

乔木：加拿大唐棣、茶条槭、美国皂荚"阴凉大师"、黑野樱、暴马丁香"象牙白丝"、金钟柏"簇穗萱草"。

灌木：狗枣猕猴桃、加拿大唐棣"芭蕾舞者"、黑果腺肋花楸"威肯"、忍冬锦带花、带翅卫矛"坎帕克塔斯"、啤酒花、圆锥绣球"独特"、五叶爬山虎、紫叶风箱果"空竹"、玫瑰"晨曲"。

多年生植物：软羽衣草、匍匐筋骨草"卡特琳巨人"、假升麻"樟子松"、拂子茅"卡尔·福斯特"、宽叶拂子茅、夏雪草、红花老鹳草、萱草"老琼"。

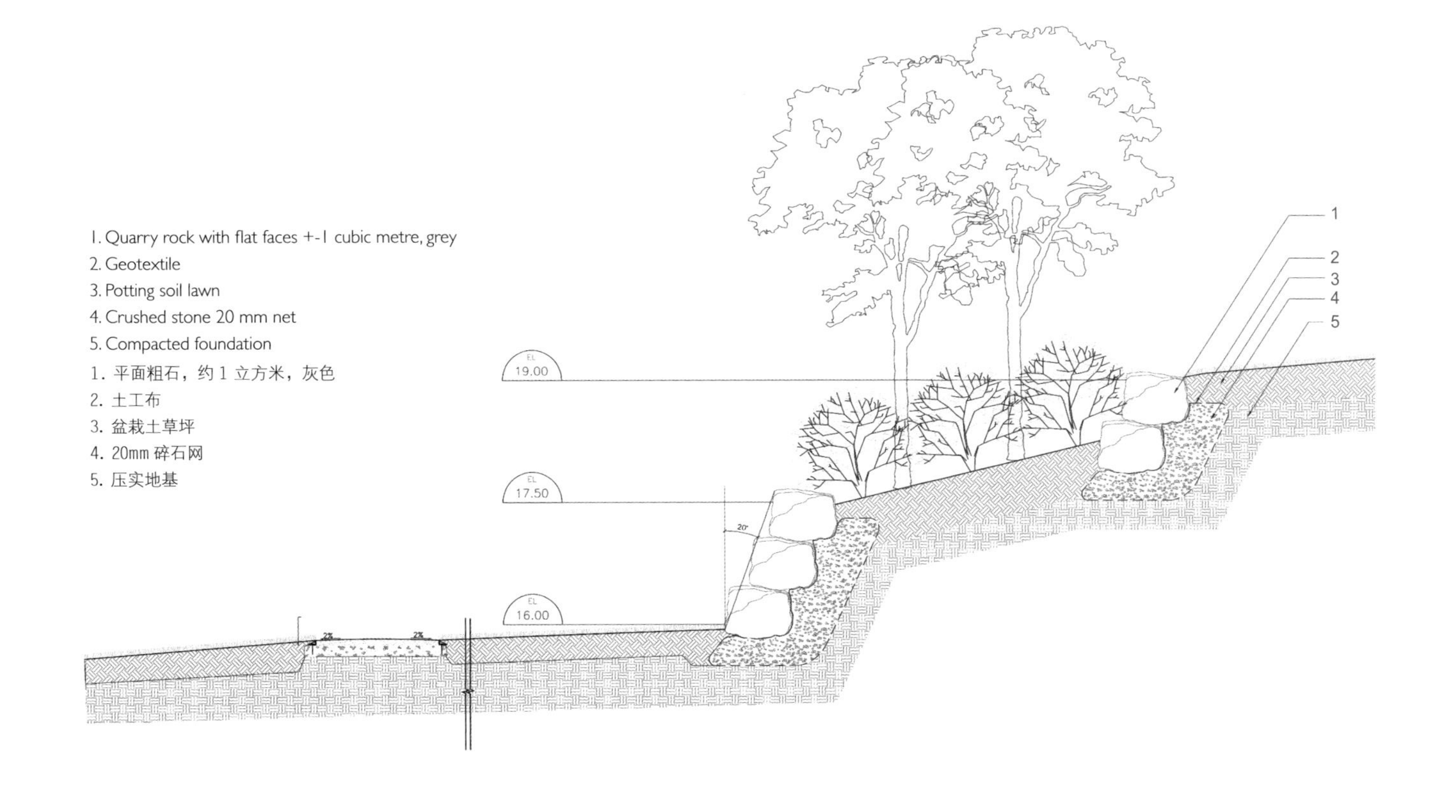

4

5

4. Formal gardens
5. Floral displays
6-7. Garden landscape brings a more agreeable view

4. 规划式庭园
5. 花卉展示
6、7. 花园式景观带来更宜人的景色

PROVINCE – ROYAL GARDEN, ZHENGZHOU

郑州信和普罗旺世——御园

Location: Zhengzhou, China
Completion: 2012
Design: Horizon & Atmosphere Landscape Co.
Photography: duo-image associates
Area: 92,205sqm

项目地点：中国，郑州
完成时间：2012年
设计师：上海翰祥景观设计咨询有限公司、瀚翔景观国际有限公司
摄影：上海柏达双影图文有限公司
面积：92,205平方米

To Enjoy Considerate and Warm Services

Throughout this over 9,000-square-metre site, each corner and each scene are refined to its best. Focused on atmosphere making, the space is simple yet elegant. It combines different pleasant dimensions into various art spaces. Take the hotel-style main entrance in the east of the site for example. The black and white ground tiles in diamond and repeated sequences reinforce the striking expression of entrance. Without paying any attention, you are already at home. Luxury and elegance, combined with visual art, raise the entrance space into a higher level. For inhabitants, returning home is an enjoyment with considerate hotel-like services, warm home and happiness of the whole community.

To Feel Walking Space

The space is arranged in multiple layers and rhythms. The whole space unfolds through the axis from the luxurious entrance, with border trees continuing the scene visually. Completed with broad central lawn and water feature wall, embellished with fragments of sculptures, the space looks artistic and unique. Walking in the project, your steps will be slowed down by the light grey solids and lush plantings. The open space on a distance scene will surprise you as you walk close. Even the murmur of stream will bring you to another dream. Walking in your way home and setting your body and soul free, you will fully enjoy the delight brought by the whole space, just like that says in the movie "Roman Holiday": "either your body or soul must be on the way".

To Taste Life and Art

The wind outside the wall whips through treetops and telling the seasonal changes of "home". The clean grey colour palette, completed with beige buildings, black water features, green sunshine lawn and seasonal plantings, consist a picturesque artistic space with the charms of black, white, grey and green. In this picture, warm afternoon sunshine penetrates the tree leaves and spills on your body. With a bench, a book and a private space, you can taste your "solitude". Otherwise with a pot of tea under the pergola, you and several friends can talk of everything under the sun.

Master plan
总平面图
1. Landscape waterscape pool
2. Fountain pool
3. Entrance tree array
4. Underground parking
5. Sunshine lawn
6. Cascading water pool
7. Leisure plaza
8. Landscape wall
9. Leisure seating
10. Lounge flower bed
11. Spraying water pool
12. Node landscape
13. Pergola lounge
1. 景观水景池
2. 大喷泉水池
3. 入口迎宾树阵
4. 地下车库
5. 阳光草坪
6. 叠水水景池
7. 休闲广场
8. 造型景墙
9. 休闲座椅
10. 休憩花坛
11. 吐水水景池
12. 节点景观
13. 花架休憩区
Provence

1

2

享 · 贴心与温馨

9,000多平方米的基地里，偌大的一个空间，每一处、每一景，精益求精。空间追求意境氛围营造，简约而不简单，糅合每个宜人的尺度，转为每一个艺术空间。就如东侧的“酒店式”小区主入口来说，黑白相间的菱形地砖，重复的序列感，视觉上突显入口立面的震撼，恍然间，身体已被带入家门。奢华、典雅，融和光影艺术，将入口空间提升到另一个境界。对居住者来说，归家就是一种享受，享受星级酒店般的贴心服务；享受家的温馨；享受整个小区所带来的幸福感。

1. Water feature
2. Sunshine lawn
3. The entrance to the community
4. Night view of the entrance

1. 景观水景池
2. 阳光草坪
3. 住区入口
4. 入口夜景

5

6

感 · 步履时空

空间布局层次富有韵律，从雍容华贵的迎宾入口区，以中轴展开，树阵让视觉产生延续性，搭配宽阔的中央草坪以及延边水景墙，残缺的雕塑小品点缀其中，空间艺术且独特。因此，步入小区所围合的场景里，淡淡的灰色硬质，配以青葱的绿色植栽，清新脱俗、怡然自得，让人脚步不由得慢了下来，仿佛时空就此停住。远方若隐若无的开放空间，看似在你的意料之中，走近时，不经意间的一个回首，又会让你“措手不及”，细细品味时，身边潺潺水声又将你带入另一种遐想。漫步于回家的林荫小道上，身心解放于天地之间，就像电影《罗马假日》里说的“身体和灵魂总要有一个在路上”，此时此景，尽情感知整个空间所带来的惬意。

品 · 生活与艺术

墙外的风，拂过树梢，穿过耳朵，诉说着“家”的四季变化。简约的灰白空间层次和色彩关系，搭配建筑本体的米黄色，辅以黑色的水景小品、绿色的阳光草坪以及随着季节变化的复层植栽，黑、白、灰、绿的魅力让整体空间呈现出如画般艺术。画面里，树阴下，休闲的

午后，暖暖的阳光透过树叶洒在身上，一条长椅，一本书，独处私密空间，享受孤独者的“孤独”。抑或许，几个老友，花架里，品一壶清茗，谈天说地，回忆起上次聚会的美好。品生活、品艺术，如此悠然自得，夫复何求?

5-6. Pergola for relaxing
7. Relaxing plaza
8. Waterscape pool

5、6. 花架休憩区
7. 休闲广场
8. 叠水水景池

MOUNTAIN VILLAS OF JINGRUI GROUP, SHAOXING

绍兴景瑞“望府”二期

Location: Shaoxing, Zhejiang
Completion: On going
Design: SED Landscape Architects Co.,LTD.
Photography: SED Landscape Architects Co.,LTD.
Area: 300,000sqm
项目地点：浙江，绍兴
完成时间：在建
设计师：SED新西林景观国际
摄影：SED新西林景观国际
面积：300,000平方米

Situating at Shaoxing, Zhejiang province, the site is part of the Keyan scenic area which proves its convenient traffic, abundant landscape resources and obvious location advantages. A splendid example about the exquisite details treatment and materials application has been set up by the surrounded high quality projects such as Keyan Golf Villas, Keyan Jiahua Fragrant Garden.

Good inheritance to the architecture style is fully shown in the landscape design which demonstrates itself as elegance and simplicity of French Neo-classic but also features as natural and ecological. Harmony of paving materials, colours and architecture materials, as well as the consistency of landscape structures, ornaments and architecture form, enables the integration of landscape and architecture.

The project site is embraced with Dongdan Mountain, with the bluff on its west side. The design brings in the unique concept of "Hidden Mountain – Rock Spa Resort Town" based on the inborn conditions of natural resources, to create a landmark ecological fitness park. The unparalleled regional features and the functions of exercise and fitness endowed by the design, compose the overall connotative tone, distinguishing itself from other projects.

The design tries to introduce a new lifestyle – "Elementary seclusion in mountains, superior seclusion in town" – to express generosity in hidden elements and luxury in low profile. Lush and picturesque landscape creates a high-quality living environment which is similar to a resort. The residents will enjoy a combination of luxurious services and peaceful comforts in this hidden heaven.

As for the unique geological environment of Dongdan Mountain, the designers make use of the topography and retain the existing vegetation, creating a mountainous club landscape and a natural living park for fitness. Through terrain transformation, the designers solve the difference of water level. They use terraced turf slope and rock band to treat the existing riverway and artificial brook, creating a hydrophilic space with consideration of standard flood level.

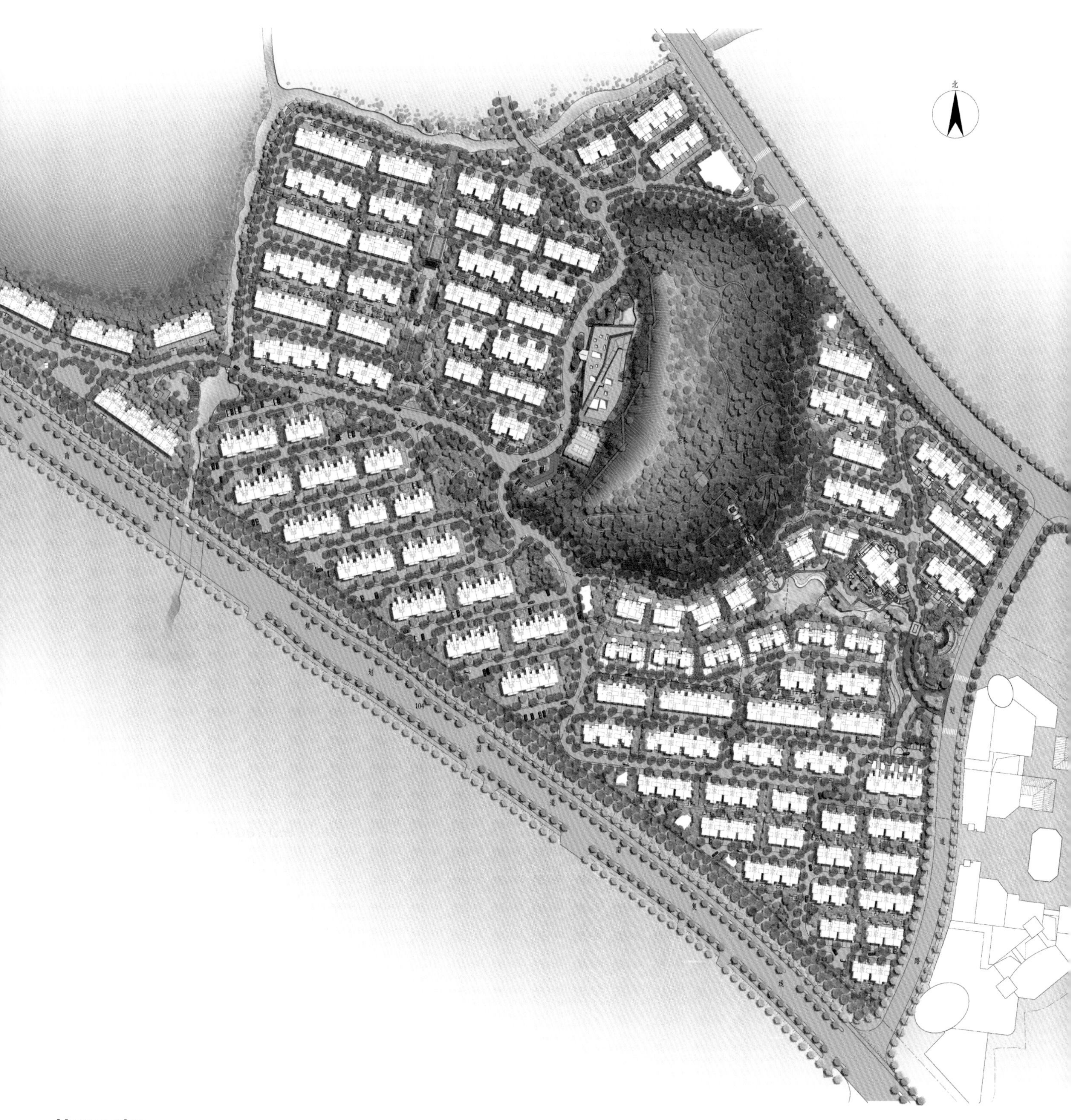

Master plan

总平面图

1. Good inheritance to the architecture style is fully shown in the landscape design
2-4. The pavement detail
5. The landscape furniture
6. The unique concept of "Hidden Mountain – Rock Spa Resort Town"

1. 景观设计延续建筑风格，体现法式新古典的简雅精辟
2～4. 铺装细节图
5. 景观小品
6. 独特的"隐山——岩SPA度假小镇"理念

5

With "Ecological Fitness" as its essence, the project selects several species of local herbs to introduce a "fitness" concept. By the configuration of herbs, fitness paths (i.e. the jogging loop which also functions as bicycle lane around main road of the living park) and mountainous landscape of fitness and leisure space carries out the "fitness" concept through the project. The humanised design reflects humanistic care and returns to the origin of fitness living.

6

项目位于浙江省绍兴县，基地地处柯岩风景区板块，紧靠东担山，其交通便利，景观资源丰富，区位优势明显。专案周边有柯岩高尔夫别墅区、柯岩嘉华馥园，均为高质量楼盘，其景观中细腻的细节处理及材料的应用都为本项目树立了很好的典范。

景观设计延续建筑风格，体现法式新古典的简雅精辟，整体偏向自然生态。景观设计中将铺装材料及色彩与建筑材质色调相协调，构筑物及小品与建筑形式相统一，使景观与建筑风格和谐共处。

“岩" SPA 生态养生公园

专案环绕东担山布置，其山体西侧为矿山断崖。设计利用自然山地的先天条件，将独特的“隐山——岩SPA度假小镇”理念引入其中，打造具有标志性的生态养生公园。得天独厚的地域性特征，加之设计赋予其健身、养生的功能，构成项目的整体内涵基调，也是其区别于其他项目的一大亮点。

设计欲表现一种全新的生活姿态——“小隐于山，大隐于市”，以隐秘显大气，以低调显奢华。营造碧翠山水，烟霞婀娜的世外桃源画境，将尊贵避世、隐于山水之间的高质量生活小区尽展眼前。在富丽堂皇与宁静安逸融汇相生的简约尊贵中，使居者感受雅郡间的归隐山林之情致。

对于东担山的特殊地质环境，我们在设计中通过保留、利用其地势地貌及现状植被，打造出极具地形特征的山地休闲会所景观和自然山地健身、养生的小区公园。其水位高差问题，我们通过地形改造，利用梯级草坡和迭石驳岸等设计手法来处理现状河道及人工水溪，在考虑到洪水位标高的同时营造出一定的亲水空间。

“生态养生”为项目的本质精髓，在植物选种时，应用了当地具有药用价值的康体植物，引入了“健康”的理念，并通过康体植物的配置、健身功能道（即围绕园区主干道设计的慢跑道、自行车道合二为一的多功能道路）以及山地健身休闲空间的景观化处理，将“康体、养生”的主题思想贯穿全园，运用人性化的设计，体现人文关怀，回归养生居住的本原。

7

8

7.The path with green plants
8.The wandering and quiet path
9.The wall with lush green plants
10.The project selects several species of local herbs to introduce a "fitness" concept
11.The stairs

7. 道路旁边种满了绿色植物
8. 蜿蜒幽静的小路
9. 墙下种满了茂盛的植物
10. 在植物选种时，应用了当地具有药用价值的康体植物
11. 台阶

VANKE VEGA BAY VILLAS, SHENZHEN

万科深圳天琴湾别墅

Location: Shenzhen, China
Completion: 2011
Design: Line and Space,LLC, SED Landscape Architects Co.,LTD.
Photography: SED Landscape Architects Co.,LTD.
Area: 300,000sqm

项目地点：中国，深圳
完成时间：2011年
设计师：Line and Space,LLC / SED新西林景观国际
摄影：SED新西林景观国际
面积：300,000平方米

Located in the resort of Shenzhen Gold Coast, between Dameisha and Xiaomeisha, Tianqin Bay villa development enjoys an excellent ocean view and luxurious supporting facilities. For this villa development in South China which boasts best of its kind in South China, "Ocean View" is its greatest landscape advantage. The landscape designers highlight the development's identity and uniqueness through this scarce resource. With the theme "Expressing noble heart and enjoying free life", the landscape design emphasises Tianqin Bay's market positioning and creates a seafront high-class villa development with harmonious landscape, unique architectural space, elegant community atmosphere and noble lifestyle. The design should both show the project's luxury to promote clients' identities and satisfy their pursuits of freedom and ocean feeling. The 48 villas each possess their individual territory, which integrate with surroundings appropriately, without any artificial feeling.

Owners who live in Tianqin Bay "see nothing but mountains, sea and sky; hear nothing but ocean billows, sound of wind and bird tweet; smell nothing but fragrance of flowers and fresh air". They "live in a retreat and enjoy bay life". The landscape designers use natural elements complemented with contemporary facilities to create a unique leisure living environment, where people enjoy sky of Miami, coast of Naples, coast road of Melbourne and resort of Hawaii without leaving home.

The site has a mountain topography, so enclosure walls are arranged in response to slope instead of in a broken line, which otherwise will increase construction difficulty. Mainly consist of clean columns and wrought iron, the walls are simple styled and smartly hidden in the environment. With walls closely integrated in natural environment, the design ensures owner's security without cutting the internal and external environment. The designers use light overhangs and minimise the use of retaining walls, which protects natural topography and planting to a maximum level. The landscape design aims to achieve a balance between the development and environment. Since the discrepancy in elevation is nearly 70 metres, vertical site landscape design becomes the focus of the work.

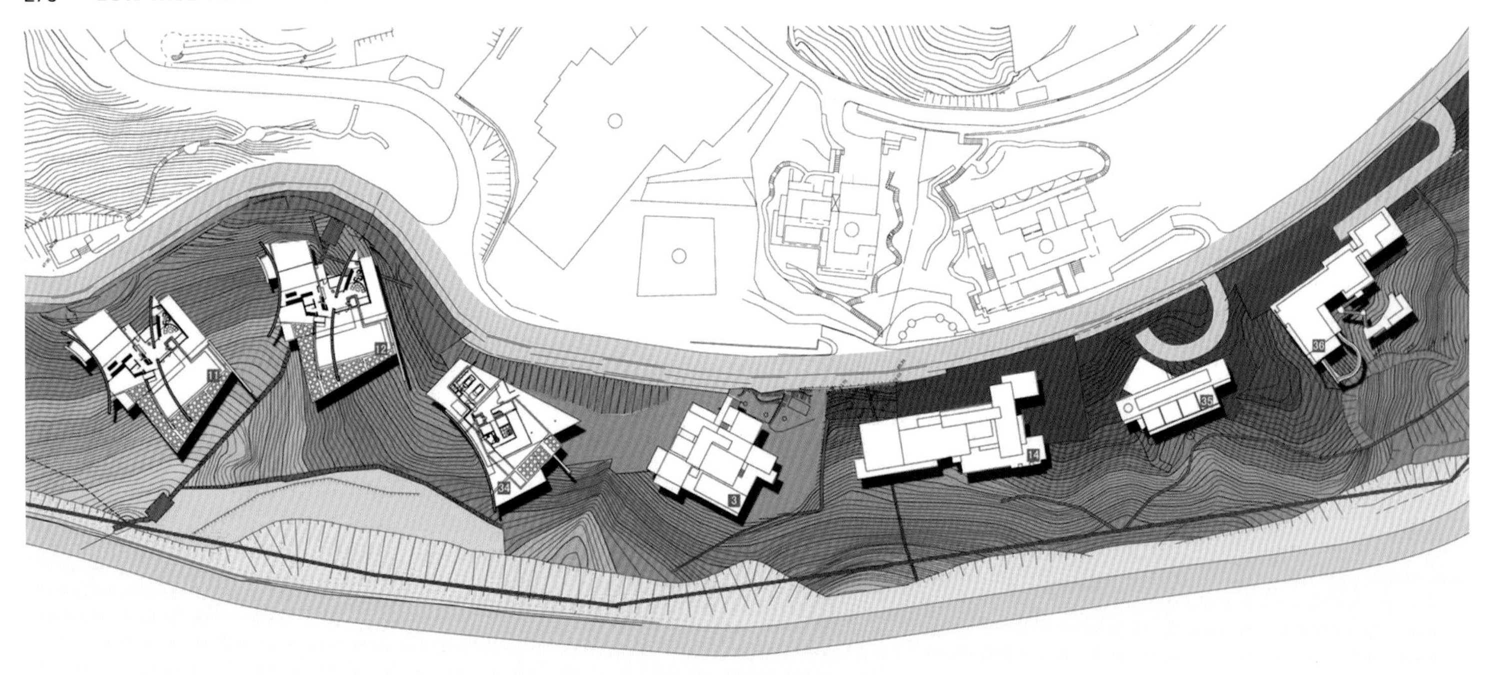

1. The site has a mountain topography
2. The path with green plants
3. Enclosure walls are arranged in response to slope instead of in a broken line
4-6. Natural elements are complemented with contemporary facilities to create a unique leisure living environment

1. 本案基地为山地地形
2. 种满绿色植物的小路
3. 围墙的设置具坡度围墙特点，顺坡度形式而设，不采用折线布局
4～6. 自然元素配合现代设施来营造一个独特的休闲居住环境

天琴湾位于深圳东部黄金海岸旅游度假胜地——大、小梅沙之间，拥有无敌海景和独一无二的至尊配套设施。作为华南区首席别墅，“海景”是项目最大景观资源优势，通过资源的稀缺性突出该别墅项目产品的差异性和唯一性。景观设计充分体现天琴湾项目的市场定位，以“演绎贵族心灵，享受自在人生”为主题，创造一个景观环境自然和谐、建筑空间独具特色、社区氛围高贵典雅、生活品质格调高尚的海滨高级别墅区，既要表现出项目的尊贵感，提升客户身份；又要满足客户对自由的追求，对海的向往。48套别墅，每套都别具一格拥有自己的独自领地，有它自身的艺术气质，与周围环境的恰如其分融合，没有人工雕琢之感。

住在这里的业主“看到的只有山、水、天；听到的只有海涛、风声、鸟鸣；闻到的只是山间的花香和空气中负离子的味道”，可以“隐居山林，享受海湾生活”。设计师运用自然元素配合现代设施来营造一个独特的休闲居住环境，让人足不出户也可以享受迈阿密的天空、那不勒斯的海岸、墨尔本的海岸公路、夏威夷的度假天堂……

3

4

5

6

本案基地为山地地形，围墙的设置具坡度围墙特点，顺坡度形式而设，不采用折线布局，以降低施工难度。围墙整体造型简洁，墙体隐蔽性强，合计元素以简洁的立柱和铁艺为主。与自然环境紧密融合，在确保安全防护的功能前提下又能使内外环境隔而不断。以轻巧的悬挑结构为主，少做挡土墙，最大程度保护现有的自然山体和绿化环境。设计力求与环境自然和谐。项目基地前后高差近70米，竖向场地景观设计成为工作重点。

7. Live in a retreat and enjoy bay life
8. The pavilion
9. Mainly consist of clean columns and wrought iron, the walls are simple styled and smartly hidden in the environment
10. Exterior space for relaxing and enjoying the sea

7. 隐居山林，享受海湾生活
8. 凉亭
9. 主要采用了简洁的立柱和铁艺，与自然环境紧密融合
10. 隐居山林，享受海湾生活

9

10

TANG ISLAND, SUZHOU

苏州棠北

Location: Suzhou, China
Completion: 2012
Design: Tsutomu Yoshizawa (Yoshiki Toda Landscape & Architect Co.,Ltd.)
Photography: Yanlord Land Co., Ltd.
Client: Yanlord Land Co., Ltd.
Area: 53,800sqm

项目地点：中国，苏州
完成时间：2012年
设计师：吉泽力（株式会社户田芳树风景计画）
摄影：仁恒地产（苏州）有限公司
客户：仁恒地产（苏州）有限公司
面积：53,800平方米

Located in a small island, the project takes broad Dushu Lake as its background. The landscape design aims to explore every features of the project and create an excellent housing environment. In the designers' interpretation, an excellent housing environment is: Sediment from Su Zhou's tradition and culture; Detachment from openness of a resort-like environment; Comfort from rich natural elements; Lifestyle from request of quality; Maturity from witness of time; Simplicity from pursuit of Zen.

To achieve this aim, the designers have taken following methods: first, they developed a detailed subdivided environment to make the project last with time; second, they created a home-returning experience with rich plots.

The former is a calm and relaxing space. Common landscape elements such as water and forest contrast with modern buildings, while they complement each other as well. A meandering creek runs through the whole site. Along our feet, stone paving and features with abundant plantings ease our heart. Although these landscape elements are common, the designers accent them with modern elements, providing the space rhythms and dynamics. Meanwhile, the Chinese traditional themes of "Plum, Orchid, Bamboo and Chrysanthemum" express a transition from urban environment to suburban environment. The latter is expressed in multi-layered home-entry experience: people travel through various custom designed entrances, go across delicate bridges, confront with terraced water features and pass through shades of tree canopies to return their homes.

The designers configured large amount of water and stone features in living space to respond to the traditions of Su Zhou, a famous watery town. These design elements not only express the respect to culture and tradition, but also enhance the intimacy of housing and highlight the owner's dignity.

Indulged in this natural, quiet and comfortable environment, with the footprints of time, your lifestyle will tend to real beauty and purity.

1. People comes to the bridge following the sound of water guegling of the water fall

1. 伴随着潺潺水声的跌水迎接着过桥而来的人们

2

2. Concise and modern door scene
3. The succinct bridge design
4. The rhythmical stone pavement of the bridge, showing respect to the culture and tradition of Suzhou
5. The water feature on the two sides of the bridge

2. 简洁、摩登的大门景象
3. 桥的设计简洁、大气
4. 桥面采用富有节奏感的石材铺装，表达了对苏州文化与传统的尊重
5. 桥两侧的水景

1. Water deck
2. Pavilion
3. South island
4. Parking apron
5. Barbecue deck
6. Connecting bridge
7. Pedestrian bridge
8. Garden swimming pool
9. Wooden bridge walkway
10. Reflection pool
11. Connecting bridge
12. Terraced deck
13. Yacht marina
14. Trickle
15. Entrance bridge of individual villa
16. Water ambulatory
17. Bank stone deck
18. Sundeck of individual villa
19. Island
20. Entrance vehicular bridge of individual villa
21. Club parking
22. Main entrance
23. Approach bridge
24. Cascading water feature
25. Wild bird conservation zone

1. 亲水平台
2. 凉亭
3. 南部小岛
4. 停机坪
5. 烧烤平台
6. 连接桥梁
7. 人行天桥
8. 庭院泳池
9. 木桥散步道
10. 水镜
11. 曲水连接桥梁
12. 阶梯式平台
13. 游艇码头
14. 细流
15. 个宅入口小桥
16. 曲水回廊
17. 岸边石台
18. 个宅内平台式晒台
19. 小岛
20. 个宅入口车行桥
21. 会所停车场
22. 入口大门
23. 引桥
24. 落水
25. 野鸟保护区

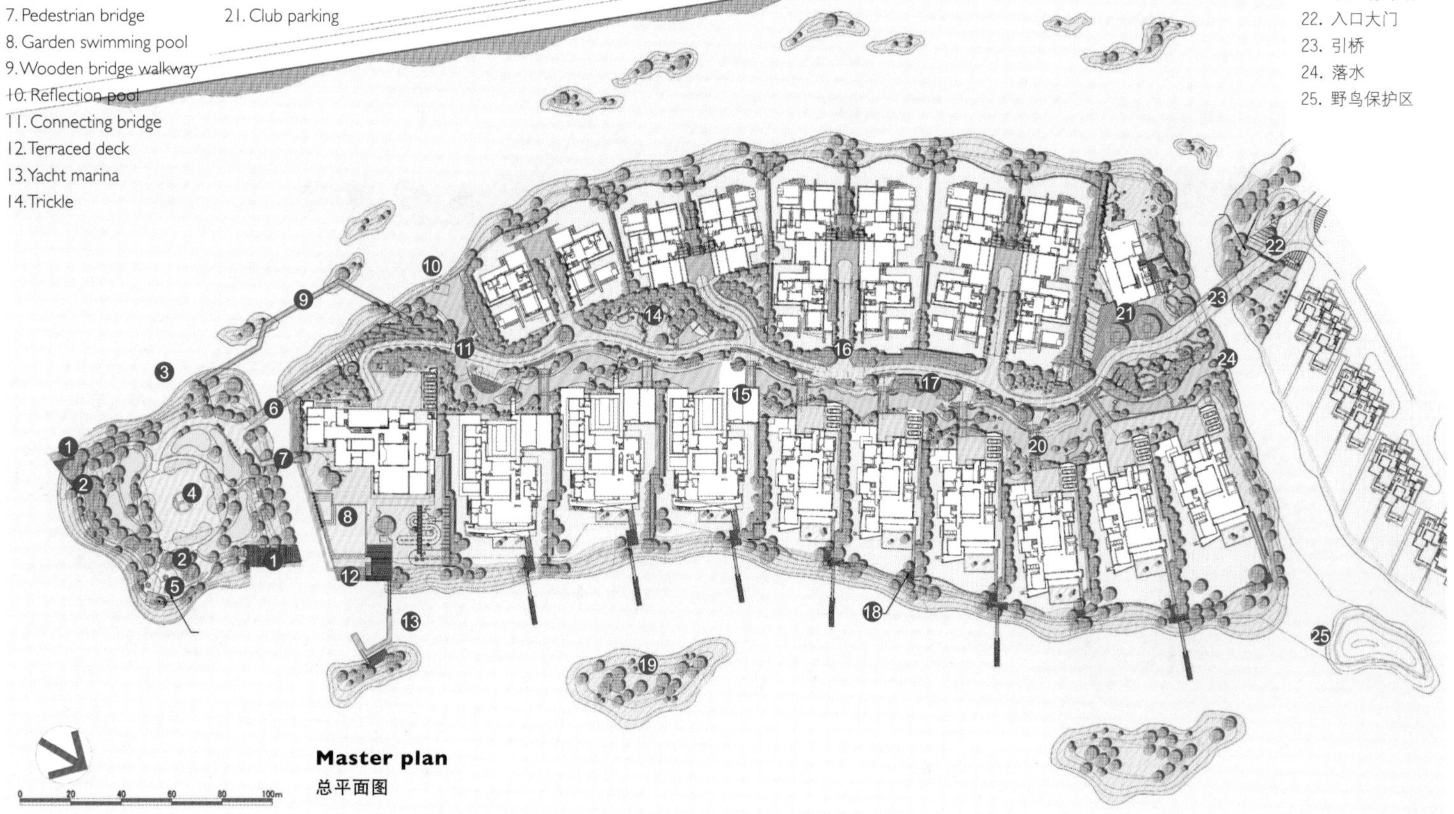

Master plan
总平面图

3

4

5

本项目位于一座小岛，以宽广的独墅湖为背景。挖掘项目具备的一切特性，营造一个极致的住宅环境，是景观所要追求的目标。何谓“极致”的生活空间？设计师对其的解释如下：是一种沉淀，源自水乡苏州的传统与文化；是一种超脱，源自充满如度假区才有的开放感；是一种恬适，源自丰富的自然；是一种格调，源自对品质的追求；是一种成熟，源自岁月的见证；是一种返璞，源自对禅意的追寻。

为了营造这样的空间，设计师采取了以下的方式：一是缔造极致细分的架构，使整个项目成为具有岁月功效的空间，让时间的流转为之增添色彩；二是让归宅的体验充满情节。

前者是一个宁静、悠然的空间：以常见的水和森林作为景观，与现代形态的建筑形成对比的同时，相互衬托；一条蜿蜒、缓缓流过的小溪贯穿用地；人们的足边空间则以石材为主，加上层次丰富的植栽，舒缓着人们的身心。虽说是常见的景观，但设计师在重点部分采用现代风格的设计，赋予空间以节奏感并使之张弛有度。与此同时，引入“梅、兰、竹、菊四

6

7

君子”作为主题，展现从“里”到“野”，即从城市过渡到郊外的空间推移。后者则通过多层次的入户体验得以体现：经过特制的入口大门、跨过精致的小桥、对望层层跌水、穿行过树冠形成的华盖来到宅邸。

设计时，注重结合水乡苏州的吴韵风情，在生活空间中配置了大量的水和石，运用了多样的手法。不仅充分表达了对文化与传统的尊重，还含蓄却不折不扣地提高了宅邸的私属性，低调地彰显了户主的尊贵感。

设计以禅的精神为基底，在空间中大量采用了诸如：摇曳、辉映、隐约、叠加、交错、引入、剪切、糅合等设计手法，演绎出多姿多彩的风景。篇头提及的开放性、丰富变化的植栽以及显现延伸感的景观，将这些要素进行叠加，不仅舒适惬意之感会油然而生，而且形成的生活空间也充满着对大自然的热爱。环绕在自然之中、恬适的空间，如同年轮，随着岁月的印痕，格调将越来越至真、至美、至纯……

6. The concise pond at the entrance which shows the contemporary architecture
7. The path changes wonderingly and slowly
8. The atrium embodies the zen mood
9. Covered deck chair, which is comfortable and relaxing; the connection of stone and wood deck reflects contemporary style

6. 入口处简洁的水庭展现现代建筑的风采
7. 蜿蜒的道路舒展地变化着、悠悠缓缓
8. 中庭体现了禅的意境
9. 有篷的躺椅，舒适惬意之感会油然而生；石材和木质甲板的衔接，体现现代风格的设计

Index 索引

C.F. Møller Architects
http://www.cfmoller.com

ONG&ONG Pte Ltd
http://ong-ong.com

ICN Design International Pte Ltd.
http://www.icn-design.com.sg

Tierra Design (S) Pte Ltd
http://www.tierradesign.com.sg

Espace Libre
http://www.espace-libre.fr

EARTHSCAPE
http://www.earthscape.co.jp

HASSEL
http://hassellstudio.com

Horizon & Atmosphere Landscape Co.
http://www.hsland.com.tw

SED Landscape Architects Co.,LTD.
http://www.sedgroup.com

One Landscape Design Limited
http://www.one-landscape.com

wa-so design
http://www. wa-so.jp

Taylor Cullity Lethlean
http://www.tcl.net.au

Williams Asselin Ackaoui & Associates Inc
http://waa-ap.com

Line and Space,LLC, SED Landscape Architects Co.,LTD.
http://www.lineandspace.com

图书在版编目（CIP）数据

居住区景观设计 / （印）程奕智编；常文心，杨莉译. -- 沈阳：辽宁科学技术出版社，2015.3
ISBN 978-7-5381-9155-4

Ⅰ. ①居… Ⅱ. ①程… ②常… ③杨… Ⅲ. ①居住区－景观设计 Ⅳ. ①TU984.12

中国版本图书馆CIP数据核字(2015)第046772号

出版发行：辽宁科学技术出版社
（地址：沈阳市和平区十一纬路29号 邮编：110003）
印 刷 者：利丰雅高印刷（深圳）有限公司
经 销 者：各地新华书店
幅面尺寸：225mm×285mm
印 张：18
插 页：4
字 数：50千字
印 数：1～1500
出版时间：2015年 3 月第 1 版
印刷时间：2015年 3 月第 1 次印刷
责任编辑：宋丹丹
封面设计：周 洁
版式设计：周 洁
责任校对：周 文
书 号：ISBN 978-7-5381-9155-4
定 价：298.00元

联系电话：024-23284360
邮购热线：024-23284502
http://www.lnkj.com.cn